Hidden Markov Models for Time Series

An Introduction Using R

MONOGRAPHS ON STATISTICS AND APPLIED PROBABILITY

General Editors

J. Fan, V. Isham, N. Keiding, T. Louis, R. L. Smith, and H. Tong

Monographs on Statistics and Applied Probability 110

Hidden Markov Models for Time Series

An Introduction Using R

Walter Zucchini
Georg–August–Universität
Göttingen, Germany

Iain L. MacDonald
University of Cape Town
South Africa

CRC Press
Taylor & Francis Group
Boca Raton   London   New York

CRC Press is an imprint of the
Taylor & Francis Group an **informa** business
A CHAPMAN & HALL BOOK

Chapman & Hall/CRC
Taylor & Francis Group
6000 Broken Sound Parkway NW, Suite 300
Boca Raton, FL 33487-2742

© 2009 by Walter Zucchini and Iain MacDonald
Chapman & Hall/CRC is an imprint of Taylor & Francis Group, an Informa business

International Standard Book Number-13: 978-1-58488-573-3 (Hardcover)

Library of Congress Cataloging-in-Publication Data

Zucchini, W.
   Hidden Markov models for time series : an introduction using R / Walter Zucchini,
Iain L. MacDonald.
      p. cm. -- (Monographs on statistics and applied probability ; 110)
   Includes bibliographical references and index.
   ISBN 978-1-58488-573-3 (hardcover : alk. paper)
   1. Time-series analysis. 2. Markov processes. 3. R (Computer program language) I.
MacDonald, Iain L. II. Title. III. Series.

QA280.Z83 2009
519.5′5--dc22   2009007294

Visit the Taylor & Francis Web site at
http://www.taylorandfrancis.com

and the CRC Press Web site at
http://www.crcpress.com

Für Hanne und Werner,
mit herzlichem Dank für Eure Unterstützung
bei der Suche nach den versteckten Ketten.

Contents

Preface

In the eleven years since the publication of our book *Hidden Markov and Other Models for Discrete-valued Time Series* it has become apparent that most of the 'other models', though undoubtedly of theoretical interest, have led to few published applications. This is in marked contrast to hidden Markov models, which are of course applicable to more than just *discrete-valued* time series. These observations have led us to write a book with different objectives.

Firstly, our emphasis is no longer principally on discrete-valued series. We have therefore removed Part One of the original text, which covered the 'other models' for such series. Our focus here is exclusively on hidden Markov models, but applied to a wide range of types of time series: continuous-valued, circular, multivariate, for instance, in addition to the types of data we previously considered, namely binary data, bounded and unbounded counts and categorical observations.

Secondly, we have attempted to make the models more accessible by illustrating how the computing environment **R** can be used to carry out the computations, e.g., for parameter estimation, model selection, model checking, decoding and forecasting. In our previous book we used proprietary software to perform numerical optimization, subject to linear constraints on the variables, for parameter estimation. We now show how one can use standard **R** functions instead. The **R** code that we used to carry out the computations for some of the applications is given, and can be applied directly in similar applications. We do not, however, supply a ready-to-use package; packages that cover 'standard' cases already exist. Rather, it is our intention to show the reader how to go about constructing and fitting application-specific variations of the standard models, variations that may not be covered in the currently available software. The programming exercises are intended to encourage readers to develop expertise in this respect.

The book is intended to illustrate the wonderful plasticity of hidden Markov models as general-purpose models for time series. We hope that readers will find it easy to devise for themselves 'customized' models that will be useful in summarizing and interpreting their data. To this end we offer a range of applications and types of data — Part Two is entirely devoted to applications. Some of the applications appeared in the original text, but these have been extended or refined.

Our intended readership is applied statisticians, students of statistics, and researchers in fields in which time series arise that are not amenable to analysis by the standard time series models such as Gaussian ARMA models. Such fields include animal behaviour, epidemiology, finance, hydrology and sociology. We have tried to write for readers who wish to acquire a general understanding of the models and their uses, and who wish to apply them. Researchers primarily interested in developing the theory of hidden Markov models are likely to be disappointed by the lack of generality of our treatment, and by the dearth of material on specific issues such as identifiability, hypothesis testing, properties of estimators and reversible jump Markov chain Monte Carlo methods. Such readers would find it more profitable to refer to alternative sources, such as Cappé, Moulines and Rydén (2005) or Ephraim and Merhav (2002). Our strategy has been to present most of the ideas by using a single running example and a simple model, the Poisson–hidden Markov model. In Chapter 8, and in Part Two of the book, we illustrate how this basic model can be progressively and variously extended and generalized.

We assume only a modest level of knowledge of probability and statistics: the reader is assumed to be familiar with the basic probability distributions such as the Poisson, normal and binomial, and with the concepts of dependence, correlation and likelihood. While we would not go as far as Lindsey (2004, p. ix) and state that 'Familiarity with classical introductory statistics courses based on point estimation, hypothesis testing, confidence intervals [...] will be a definite handicap', we hope that extensive knowledge of such matters will not prove necessary. No prior knowledge of Markov chains is assumed, although our coverage is brief enough that readers may wish to supplement our treatment by reading the relevant parts of a book such as Grimmett and Stirzaker (2001). We have also included exercises of a theoretical nature in many of the chapters, both to fill in the details and to illustrate some of the concepts introduced in the text. All the datasets analysed in this book can be accessed at the following address: `http://134.76.173.220/hmm-with-r/data`.

This book contains some material which has not previously been published, either by ourselves or (to the best of our knowledge) by others. If we have anywhere failed to make appropriate acknowledgement of the work of others, or misquoted their work in any way, we would be grateful if the reader would draw it to our attention. The applications described in Chapters 14, 15 and 16 contain material which first appeared in (respectively) the *South African Statistical Journal*, the *International Journal of Epidemiology* and *Biometrics*. We are grateful to the editors of these journals for allowing us to reuse such material.

We wish to thank the following researchers for giving us access to their data, and in some cases spending much time discussing it with us: David Bowie, Graham Fick, Linda Haines, Len Lerer, Frikkie Potgieter, David Raubenheimer and Max Suster.

We are especially indebted to Andreas Schlegel and Jan Bulla for their important inputs, particularly in the early stages of the project; to Christian Gläser, Oleg Nenadić and Daniel Adler, for contributing their computing expertise; and to Antony Unwin and Ellis Pender for their constructive comments on and criticisms of different aspects of our work. The second author wishes to thank the Institute for Statistics and Econometrics of Georg–August–Universität, Göttingen, for welcoming him on many visits and placing facilities at his disposal. Finally, we are most grateful to our colleague and friend of many years, Linda Haines, whose criticism has been invaluable in improving this book.

Göttingen
November 2008

Notation and abbreviations

Since the underlying mathematical ideas are the important quantities, no notation should be adhered to slavishly. It is all a question of who is master.

Bellman (1960, p. 82)

[...] many writers have acted as though they believe that the success of the Box–Jenkins models is largely due to the use of the acronyms.

Granger (1982)

Notation

Although notation is defined as it is introduced, it may also be helpful to list here the most common meanings of symbols, and the pages on which they are introduced. Matrices and vectors are denoted by bold type. Transposition of matrices and vectors is indicated by the prime symbol: $'$. All vectors are row vectors unless indicated otherwise.

Symbol	Meaning	Page
$\mathbf{A}(,i)$	ith column of any matrix $\mathbf{A}$	86
$A_n(\kappa)$	$I_n(\kappa)/I_0(\kappa)$	160
$\mathbf{B}_t$	$\boldsymbol{\Gamma}\mathbf{P}(x_t)$	37
C_t	state occupied by Markov chain at time t	16
$\mathbf{C}^{(t)}$	$(C_1, C_2, \ldots, C_t)$	16
$\{g_t\}$	parameter process of a stochastic volatility model	190
I_n	modified Bessel function of the first kind of order n	156
l	log-likelihood	21
L or L_T	likelihood	21, 35
$\log$	logarithm to the base e	
m	number of states in a Markov chain,	17
	or number of components in a mixture	7
$\mathbb{N}$	the set of all positive integers	
N_t	nutrient level	220
$N(\bullet; \mu, \sigma^2)$	distribution function of general normal distribution	191
$n(\bullet; \mu, \sigma^2)$	density of general normal distribution	191
p_i	probability mass or density function in state i	31
$\mathbf{P}(x)$	diagonal matrix with ith diagonal element $p_i(x)$	32
$\mathbb{R}$	the set of all real numbers	

T	length of a time series	35
$\mathbf{U}$	square matrix with all elements equal to 1	19
$\mathbf{u}(t)$	vector $(\Pr(C_t = 1), \ldots, \Pr(C_t = m))$	17
$u_i(t)$	$\Pr(C_t = i)$, i.e. ith element of $\mathbf{u}(t)$	32
w_t	$\boldsymbol{\alpha}_t \mathbf{1}' = \sum_i \alpha_t(i)$	46
X_t	observation at time t, or just tth observation	30
$\mathbf{X}^{(t)}$	$(X_1, X_2, \ldots, X_t)$	30
$\mathbf{X}^{(-t)}$	$(X_1, \ldots, X_{t-1}, X_{t+1}, \ldots X_T)$	76
$\mathbf{X}_a^b$	$(X_a, X_{a+1}, \ldots, X_b)$	61
$\boldsymbol{\alpha}_t$	(row) vector of forward probabilities	38
$\alpha_t(i)$	forward probability, i.e. $\Pr(\mathbf{X}^{(t)} = \mathbf{x}^{(t)}, C_t = i)$	59
$\boldsymbol{\beta}_t$	(row) vector of backward probabilities	60
$\beta_t(i)$	backward probability, i.e. $\Pr(\mathbf{X}_{t+1}^T = \mathbf{x}_{t+1}^T \mid C_t = i)$	60
$\boldsymbol{\Gamma}$	transition probability matrix of Markov chain	17
γ_{ij}	(i, j) element of $\boldsymbol{\Gamma}$; probability of transition from state i to state j in a Markov chain	17
$\boldsymbol{\delta}$	stationary or initial distribution of Markov chain, or vector of mixing probabilities	18 7
$\boldsymbol{\phi}_t$	vector of forward probabilities, normalized to have sum equal to 1, i.e. $\boldsymbol{\alpha}_t / w_t$	46
Φ	distribution function of standard normal distribution	
$\mathbf{1}$	(row) vector of ones	19

Abbreviations

ACF	autocorrelation function
AIC	Akaike's information criterion
BIC	Bayesian information criterion
CDLL	complete-data log-likelihood
c.o.d.	change of direction
c.v.	coefficient of variation
HM	hidden Markov
HMM	hidden Markov model
MC	Markov chain
MCMC	Markov chain Monte Carlo
ML	maximum likelihood
MLE	maximum likelihood estimator or estimate
PACF	partial autocorrelation function
qq-plot	quantile-quantile plot
SV	stochastic volatility
t.p.m.	transition probability matrix

PART ONE

Model structure, properties and methods

CHAPTER 1

Preliminaries: mixtures and Markov chains

1.1 Introduction

Hidden Markov models (HMMs) are models in which the distribution that generates an observation depends on the state of an underlying and unobserved Markov process. They show promise as flexible general-purpose models for univariate and multivariate time series, especially for discrete-valued series, including categorical series and series of counts.

The purposes of this chapter are to provide a very brief and informal introduction to HMMs, and to their many potential uses, and then to discuss two topics that will be fundamental in understanding the structure of such models. In Section 1.2 we give an account of (finite) mixture distributions, because the marginal distribution of a hidden Markov model is a mixture distribution, and then, in Section 1.3, we introduce Markov chains, which provide the underlying 'parameter process' of a hidden Markov model.

Consider, as an example, the series of annual counts of major earthquakes (i.e. magnitude 7 and above) for the years 1900–2006, both inclusive, displayed in Table 1.1 and Figure 1.1.* For this series, the application of standard models such as autoregressive moving-average (ARMA) models would be inappropriate, because such models are based on the normal distribution. Instead, the usual model for unbounded counts is the Poisson distribution, but as will be demonstrated later, the series displays considerable overdispersion relative to the Poisson distribution, and strong positive serial dependence. A model consisting of independent Poisson random variables would therefore for two reasons also be inappropriate. An examination of Figure 1.1 suggests that there may be some periods with a low rate of earthquakes, and some with a relatively high rate. HMMs, which allow the probability distribution of each observation to depend on the unobserved (or 'hidden') state of a Markov chain, can accommodate both overdispersion and serial dependence. We

---

* These data were downloaded from http://neic.usgs.gov/neis/eqlists on 25 July 2007. Note, however, that the U.S. Geological Survey has undertaken a systematic review, which is expected to lead to revision of the observations for years prior to 1990.

3

Table 1.1 *Number of major earthquakes (magnitude 7 or greater) in the world, 1900–2006; to be read across rows.*

13	14	8	10	16	26	32	27	18	32	36	24	22	23	22	18	25	21	21	14
8	11	14	23	18	17	19	20	22	19	13	26	13	14	22	24	21	22	26	21
23	24	27	41	31	27	35	26	28	36	39	21	17	22	17	19	15	34	10	15
22	18	15	20	15	22	19	16	30	27	29	23	20	16	21	21	25	16	18	15
18	14	10	15	8	15	6	11	8	7	18	16	13	12	13	20	15	16	12	18
15	16	13	15	16	11	11													

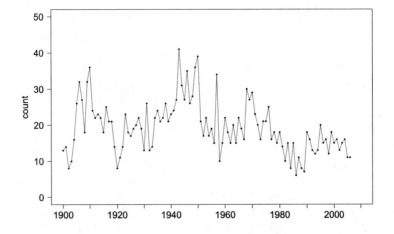

Figure 1.1 *Number of major earthquakes (magnitude 7 or greater) in the world, 1900–2006.*

shall use this series of earthquake counts as a running example in Part One of the book, in order to illustrate the fitting of a Poisson–HMM and many other aspects of that model.

Hidden Markov models have been used for at least three decades in signal-processing applications, especially in the context of automatic speech recognition, but interest in their theory and application has expanded to other fields, e.g.:

- all kinds of recognition: face, gesture, handwriting, signature;
- bioinformatics: biological sequence analysis;
- environment: wind direction, rainfall, earthquakes;

- finance: series of daily returns;

- biophysics: ion channel modelling.

Attractive features of HMMs include their simplicity, their general mathematical tractability, and specifically the fact that the likelihood is relatively straightforward to compute. The main aim of this book is to illustrate how HMMs can be used as general-purpose models for time series.

Following this preliminary chapter, the book introduces what we shall call the **basic HMM**: basic in the sense that it is univariate, is based on a homogeneous Markov chain, and has neither trend nor seasonal variation. The observations may be either discrete- or continuous-valued, but we initially ignore information that may be available on covariates. We focus on the following issues:

- parameter estimation (Chapters 3 and 4);

- point and interval forecasting (Chapter 5);

- decoding, i.e. estimating the sequence of hidden states (Chapter 5);

- model selection, model checking and outlier detection (Chapter 6).

We give one example of the Bayesian approach to inference (Chapter 7).

In Chapter 8 we discuss the many possible extensions of the basic HMM to a wider range of models. These include HMMs for series with trend and seasonal variation, methods to include covariate information from other time series, and multivariate models of various types.

Part Two of the book offers fairly detailed applications of HMMs to time series arising in a variety of subject areas. These are intended to illustrate the theory covered in Part One, and also to demonstrate the versatility of HMMs. Indeed, so great is the variety of HMMs that it is hard to imagine this variety being exhaustively covered by any single software package. In some applications the model needs to accommodate some special features of the time series. In such cases it is necessary to write one's own code. We have found the computing environment **R** (Ihaka and Gentleman, 1996; R Development Core Team, 2008) to be particularly convenient for this purpose.

Many of the chapters contain exercises, some theoretical and some practical. Because one always learns more about models by applying them in practice, and because some aspects of the theory of HMMs are covered only in these exercises, we regard these as an important part of the book. As regards the practical exercises, our strategy has been to give examples of **R** functions for some important but simple cases, and to encourage readers to learn to write their own code, initially just by modifying the functions given in Appendix A.

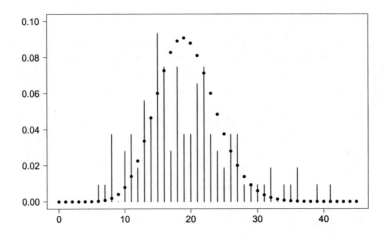

Figure 1.2 *Major earthquakes 1900–2006: bar plot of observed counts, and fitted Poisson distribution.*

1.2 Independent mixture models

1.2.1 Definition and properties

Consider again the series of earthquake counts displayed in Figure 1.1. A standard model for unbounded counts is the Poisson distribution, with its probability function $p(x) = e^{-\lambda}\lambda^x/x!$ and the property that the variance equals the mean. However, for the earthquakes series the sample variance, $s^2 \approx 52$, is much larger than the sample mean, $\bar{x} \approx 19$, which indicates strong overdispersion relative to the Poisson distribution and the inappropriateness of that distribution as a model. The lack of fit is confirmed by Figure 1.2, which displays both a bar plot of the observed counts and the fitted Poisson distribution.

One method of dealing with overdispersed observations with a bi-modal or — more generally — multimodal distribution is to use a mixture model. Mixture models are designed to accommodate unobserved heterogeneity in the population; that is, the population may consist of unobserved groups, each having a distinct distribution for the observed variable.

Consider for example the distribution of the number, X, of packets of cigarettes bought by the customers of a supermarket. The customers can be divided into groups, e.g. nonsmokers, occasional smokers, and regular smokers. Now even if the number of packets bought by customers within each group were Poisson-distributed, the distribution of X would not be Poisson; it would be overdispersed relative to the Poisson, and maybe even multimodal.

Analogously, suppose that each count in the earthquakes series is generated by one of two Poisson distributions, with means λ_1 and λ_2, where the choice of mean is determined by some other random mechanism which we call the **parameter process**. Suppose also that λ_1 is selected with probability δ_1 and λ_2 with probability $\delta_2 = 1 - \delta_1$. We shall see later in this chapter that the variance of the resulting distribution exceeds the mean by $\delta_1 \delta_2 (\lambda_1 - \lambda_2)^2$. If the parameter process is a series of independent random variables, the counts are also independent; hence the term 'independent mixture'.

In general, an independent mixture distribution consists of a finite number, say m, of component distributions and a 'mixing distribution' which selects from these components. The component distributions may be either discrete or continuous. In the case of two components, the mixture distribution depends on two probability or density functions:

component	1	2
probability or density function	$p_1(x)$	$p_2(x)$.

To specify the component, one needs a discrete random variable C which performs the mixing:

$$C = \begin{cases} 1 \text{ with probability } \delta_1 \\ 2 \text{ with probability } \delta_2 = 1 - \delta_1. \end{cases}$$

The structure of that process for the case of two continuous component distributions is illustrated in Figure 1.3. In that example one can think of C as the outcome of tossing a coin with probability 0.75 of 'heads': if the outcome is 'heads', then $C = 1$ and an observation is drawn from p_1; if it is 'tails', then $C = 2$ and an observation is drawn from p_2. We suppose that we do not know the value C, i.e. which of p_1 or p_2 was active when the observation was generated.

The extension to m components is straightforward. Let δ_1, ..., δ_m denote the probabilities assigned to the different components, and let p_1, ..., p_m denote their probability or density functions. Let X denote the random variable which has the mixture distribution. It is easy to show that the probability or density function of X is given by

$$p(x) = \sum_{i=1}^{m} \delta_i p_i(x).$$

For the discrete case this follows immediately from

$$\Pr(X = x) = \sum_{i=1}^{m} \Pr(X = x \mid C = i) \Pr(C = i).$$

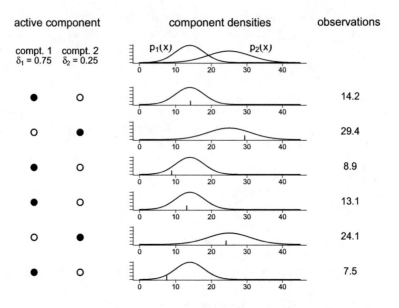

Figure 1.3 *Process structure of a two-component mixture distribution. From top to bottom, the states are 1,2,1,1,2,1. The corresponding component distributions are shown in the middle. The observations are generated from the active component density.*

Moreover, the expectation of the mixture can be given in terms of the expectations of the component distributions. Letting Y_i denote the random variable with probability function p_i, we have

$$\mathrm{E}(X) = \sum_{i=1}^{m} \mathrm{Pr}(C = i)\,\mathrm{E}(X \mid C = i) = \sum_{i=1}^{m} \delta_i\,\mathrm{E}(Y_i).$$

The same result holds, of course, for a mixture of continuous distributions.

More generally, for a mixture the kth moment about the origin is simply a linear combination of the kth moments of its components Y_i:

$$\mathrm{E}(X^k) = \sum_{i=1}^{m} \delta_i\,\mathrm{E}(Y_i^k), \quad k = 1, 2, \dots .$$

Note that the analogous result does not hold for central moments. In particular, the variance of X is not a linear combination of the variances of its components Y_i. Exercise 1 asks the reader to prove that, in the two-component case, the variance of the mixture is given by

$$\mathrm{Var}(X) = \delta_1\,\mathrm{Var}(Y_1) + \delta_2\,\mathrm{Var}(Y_2) + \delta_1\delta_2\big(\mathrm{E}(Y_1) - \mathrm{E}(Y_2)\big)^2.$$

1.2.2 Parameter estimation

The estimation of the parameters of a mixture distribution is often performed by maximum likelihood (ML). In general, the likelihood of a mixture model with m components is given, for both discrete and continuous cases, by

$$L(\boldsymbol{\theta}_1,\ldots,\boldsymbol{\theta}_m,\delta_1,\ldots,\delta_m \mid x_1,\ldots,x_n) = \prod_{j=1}^{n}\sum_{i=1}^{m}\delta_i p_i(x_j,\boldsymbol{\theta}_i). \qquad (1.1)$$

Here $\boldsymbol{\theta}_1,\ldots,\boldsymbol{\theta}_m$ are the parameter vectors of the component distributions, $\delta_1,\ldots,\delta_m$ are the mixing parameters, totalling 1, and $x_1,\ldots,x_n$ are the n observations. Thus, in the case of component distributions each specified by one parameter, $2m-1$ independent parameters have to be estimated. The maximization of this likelihood is not trivial since it is not possible to solve the maximization problem analytically. This will be demonstrated by considering the case of a mixture of Poisson distributions.

Suppose that $m=2$ and the two components are Poisson-distributed with means λ_1 and λ_2. Let δ_1 and δ_2 be the mixing parameters (with $\delta_1 + \delta_2 = 1$). The mixture distribution $p(x)$ is then given by

$$p(x) = \delta_1 \frac{\lambda_1^x e^{-\lambda_1}}{x!} + \delta_2 \frac{\lambda_2^x e^{-\lambda_2}}{x!}.$$

Since $\delta_2 = 1 - \delta_1$, there are only three parameters to be estimated: λ_1, λ_2 and δ_1. The likelihood is

$$L(\lambda_1,\lambda_2,\delta_1 \mid x_1,\ldots,x_n) = \prod_{i=1}^{n}\left(\delta_1 \frac{\lambda_1^{x_i} e^{-\lambda_1}}{x_i!} + (1-\delta_1)\frac{\lambda_2^{x_i} e^{-\lambda_2}}{x_i!}\right).$$

The analytic maximization of L with respect to λ_1, λ_2 and δ_1 would be awkward, as L is the product of n factors, each of which is a sum. First taking the logarithm and then differentiating does not greatly simplify matters, either. Therefore parameter estimation is more conveniently carried out by numerical maximization of the likelihood (or its logarithm), although the EM algorithm is a commonly used alternative: see e.g. McLachlan and Peel (2000) or Frühwirth-Schnatter (2006). (We shall in Chapter 4 discuss the EM algorithm more fully in the context of the estimation of hidden Markov models.) A useful **R** package for estimation in mixture models is `flexmix` (Leisch, 2004). However, it is straightforward to write one's own **R** code to evaluate, and then maximize, mixture likelihoods in simple cases. For instance, the following single **R** expression will evaluate the log of the likelihood of observations x (as given by Equation (1.1)) of a mixture of Poisson-distributed components, `lambda` being the vector of means and `delta` the vector containing the corresponding mixing distribution:

$$\texttt{sum(log(outer(x,lambda,dpois)\%*\%delta))}.$$

This log-likelihood can then be maximized by using (e.g.) the **R** function `nlm`. However, the parameters $\boldsymbol{\delta}$ and $\boldsymbol{\lambda}$ are constrained by $\sum_{i=1}^{m} \delta_i = 1$ and (for $i = 1, \dots, m$) $\delta_i > 0$ and $\lambda_i > 0$. It is therefore necessary to reparametrize if one wishes to use an unconstrained optimizer such as `nlm`. One possibility is to maximize the likelihood with respect to the $2m - 1$ unconstrained parameters

$$\eta_i = \log \lambda_i \quad (i = 1, \dots, m)$$

and

$$\tau_i = \log \left(\frac{\delta_i}{1 - \sum_{j=2}^{m} \delta_j} \right) \quad (i = 2, \dots, m).$$

One recovers the original parameters via

$$\lambda_i = e^{\eta_i}, \quad (i = 1, \dots, m),$$

$$\delta_i = \frac{e^{\tau_i}}{1 + \sum_{j=2}^{m} e^{\tau_i}} \quad (i = 2, \dots, m),$$

and $\delta_1 = 1 - \sum_{j=2}^{m} \delta_i$. See Exercise 3.

1.2.3 Unbounded likelihood in mixtures

There is one aspect of mixtures of continuous distributions that differs from the discrete case and is worth highlighting. It is this: it can happen that, in the vicinity of certain parameter combinations, the likelihood is unbounded. For instance, in the case of a mixture of normal distributions, the likelihood becomes arbitrarily large if one sets a component mean equal to one of the observations and allows the corresponding variance to tend to zero. The problem has been extensively discussed in the literature on mixture models, and there are those who would say that, if the likelihood is thus unbounded, the maximum likelihood estimates simply 'do not exist': see for instance Scholz (2006, p. 4630). The source of the problem, however, is just the use of densities rather than probabilities in the likelihood; it would not arise if one were to replace each density value in the likelihood by the probability of the interval corresponding to the recorded value. (For example, an observation recorded as '12.4' is associated with the interval $[12.35, 12.45)$.) One then replaces the expression $\prod_{j=1}^{n} \sum_{i=1}^{m} \delta_i p_i(x_j, \boldsymbol{\theta}_i)$ for the likelihood — see Equation (1.1) — by the **discrete likelihood**

$$L = \prod_{j=1}^{n} \sum_{i=1}^{m} \delta_i \int_{a_j}^{b_j} p_i(x, \boldsymbol{\theta}_i) \, dx, \tag{1.2}$$

Table 1.2 *Poisson independent mixture models fitted to the earthquakes series. The number of components is m, the mixing probabilities are denoted by δ_i, and the component means by λ_i. The maximized likelihood is L.*

model	i	δ_i	λ_i	$-\log L$	mean	variance
$m = 1$	1	1.000	19.364	391.9189	19.364	19.364
$m = 2$	1	0.676	15.777	360.3690	19.364	46.182
	2	0.324	26.840			
$m = 3$	1	0.278	12.736	356.8489	19.364	51.170
	2	0.593	19.785			
	3	0.130	31.629			
$m = 4$	1	0.093	10.584	356.7337	19.364	51.638
	2	0.354	15.528			
	3	0.437	20.969			
	4	0.116	32.079			
observations					19.364	51.573

where the interval (a_j, b_j) consists of precisely those values which, if observed, would be recorded as x_j. This simply amounts to acknowledging explicitly the interval nature of all 'continuous' observations. Another way of avoiding the problem would be to impose a lower bound on the variances and search for the best local maximum subject to that bound. It can happen, though, that one is fortunate enough to avoid the likelihood 'spikes' when searching for a local maximum; in this respect good starting values can help.

The problem of unbounded likelihood does not arise for discrete-valued observations because the likelihood is in that case a probability and thereby bounded by 0 and 1.

1.2.4 Examples of fitted mixture models

Mixtures of Poisson distributions

If one uses nlm to fit a mixture of m Poisson distributions ($m = 1$, 2, 3, 4) to the earthquakes data, one obtains the results displayed in Table 1.2. Notice, for instance, that the likelihood improvement resulting

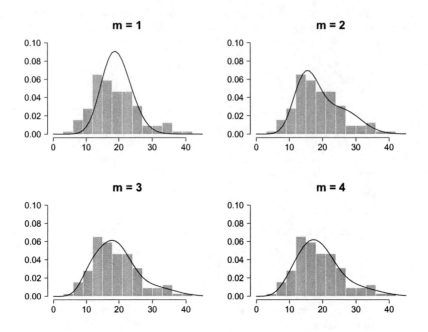

Figure 1.4 *Earthquakes data: histogram of counts, compared to mixtures of 1–4 Poisson distributions.*

from the addition of a fourth component is very small, and apparently insufficient to justify the additional two parameters. Section 6.1 will discuss the model selection problem in more detail. Figure 1.4 presents a histogram of the observed counts and the four models fitted. It is clear that the mixtures fit the observations much better than does a single Poisson distribution, and visually the three- and four-state models seem adequate. The better fit of the mixtures is also evident from the means and variances of the four models as presented in Table 1.2. In computing the means and variances of the models we have used $E(X) = \sum_i \delta_i \lambda_i$ and $\text{Var}(X) = E(X^2) - (E(X))^2$, with $E(X^2) = \sum_i \delta_i (\lambda_i + \lambda_i^2)$. For comparison we also used the **R** package `flexmix` to fit the same four models. The results corresponded closely except in the case of the four-component model, where the highest likelihood value found by `flexmix` was slightly inferior to that found by `nlm`.

Note, however, that the above discussion ignores the possibility of serial dependence in the earthquakes data, a point we shall take up in Chapter 2.

Table 1.3 *Data of Hastie* et al. *(2001), plus two mixture models. The first model was fitted by direct numerical maximization in* **R**, *the second is the model fitted by EM by Hastie* et al.

−0.39	0.12	0.94	1.67	1.76	2.44	3.72	4.28	4.92	5.53
0.06	0.48	1.01	1.68	1.80	3.25	4.12	4.60	5.28	6.22

i	δ_i	μ_i	σ_i^2	$-\log L$
1	0.5546	4.656	0.8188	38.9134
2	0.4454	1.083	0.8114	
1	0.546	4.62	0.87	(38.924)
2	0.454	1.06	0.88	

Mixtures of normal distributions

As a very simple example of the fitting of an independent mixture of normal distributions, consider the data presented in Table 8.1 of Hastie, Tibshirani and Friedman (2001, p. 237); see our Table 1.3. Hastie *et al.* use the EM algorithm to fit a mixture model with two normal components.

Our two-component model, fitted by direct numerical maximization of the log-likelihood in **R**, has log-likelihood −38.9134, and is also displayed in Table 1.3. (Here we use as the likelihood the joint density of the observations, not the discrete likelihood.) The parameter estimates are close to those quoted by Hastie *et al.*, but not identical, and by our computations their model produces a log-likelihood of approximately −38.924, i.e. marginally inferior to that of our model.

As a second example of the fitting of an independent mixture of normal distributions we now consider the durations series relating to the much-studied 'Old Faithful' geyser. We shall say more about this and a related series in Chapter 10, in particular as regards serial dependence, but for the moment we ignore such dependence and describe only the independent mixture model for the durations of eruptions. This example provides an illustration of what can go wrong if one seeks to maximize the joint density, and demonstrates how the technique of replacing 'continuous' observations by intervals is very flexible and can cope quite generally with interval censoring.

The series we consider here is taken from Azzalini and Bowman (1990), 299 observations of the durations of successive eruptions of the geyser, recorded to the nearest second except that some of the observations

Table 1.4 *Normal independent mixture models fitted to the Old Faithful dura-*
tions series by maximizing the discrete likelihood. The number of components
is m, the mixing probabilities are denoted by δ_i, and the component means and
standard deviations by μ_i and σ_i. The maximized likelihood is L.

model	i	δ_i	μ_i	σ_i	$-\log L$
$m = 2$	1	0.354	1.981	0.311	1230.920
	2	0.646	4.356	0.401	
$m = 3$	1	0.277	1.890	0.134	1203.872
	2	0.116	2.722	0.747	
	3	0.607	4.405	0.347	
$m = 4$	1	0.272	1.889	0.132	1203.636
	2	0.104	2.532	0.654	
	3	0.306	4.321	0.429	
	4	0.318	4.449	0.282	

were identified only as short (S), medium (M) or long (L); the accu-
rately recorded durations range from 0:50 (minutes:seconds) to 5:27.
There were in all 20 short eruptions, 2 medium and 47 long. For certain
purposes Azzalini and Bowman represented the codes S, M, L by durations of 2, 3, and 4 minutes' length, and that is the form in which the
data have been made available in the **R** package MASS; we refer to those
data as being in the MASS form.

When Azzalini and Bowman dichotomize the series, they do so at 3
minutes. We therefore treat S as lying below 3:00 (minutes:seconds), L as
lying above 3:00, and M, somewhat arbitrarily, as lying between 2:30 and
3:30. In the model to be discussed in Section 10.4 it will be convenient
to set a finite upper limit to the long durations; both there and here
we take 'above 3' to mean 'between 3 and 20'. We evaluate the discrete
likelihood by using Equation (1.2), with the interval (a_j, b_j) equal to (0,
3), (2.50, 3.50) and (3, 20) for S, M and L respectively, and otherwise
equal to the observation $\pm \frac{1}{2}$ second.

Table 1.4 displays the three models that were fitted by using nlm
to maximize the discrete log-likelihood, and Figure 1.5 compares them
(for $m = 2$ and 3 only) with those fitted by maximizing the continuous
likelihood of the data in the MASS form (where by 'continuous likelihood'
we mean the joint density). The attempt to fit a four-component model
by using nlm to maximize the continuous (log-) likelihood failed in that

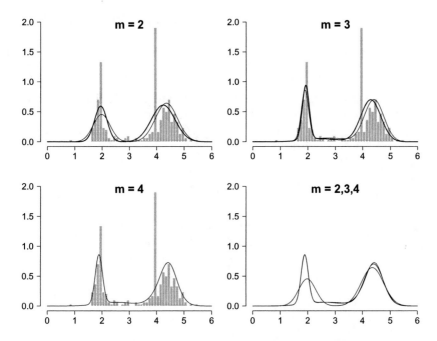

Figure 1.5 *Old Faithful durations: histogram of observations (with S,M,L re-placed by 2, 3, 4), compared to independent mixtures of 2–4 normal distribu-tions. Thick lines (only for m = 2 and 3): p.d.f. of model based on continuous likelihood. Thin lines (all cases): p.d.f. of model based on discrete likelihood.*

it terminated at a very high likelihood value, with one of the components being normal with mean 4.000 and standard deviation 1.2×10^{-6}. This is an example of the phenomenon of unbounded likelihood described in Section 1.2.3, and demonstrates the value of using the discrete likelihood rather than the continuous.

1.3 Markov chains

We now introduce Markov chains, a second building-block of hidden Mar-kov models. Our treatment is restricted to those few aspects of discrete-time Markov chains that we need. Thus, although we shall make passing reference to properties such as irreducibility and aperiodicity, we shall not dwell on such technical issues. For an excellent general account of the topic, see Grimmett and Stirzaker (2001, Chapter 6), or Feller's classic text (Feller, 1968).

1.3.1 Definitions and example

A sequence of discrete random variables $\{C_t : t \in \mathbb{N}\}$ is said to be a (discrete-time) **Markov chain** (MC) if for all $t \in \mathbb{N}$ it satisfies the **Markov property**

$$\Pr(C_{t+1} \mid C_t, \ldots, C_1) = \Pr(C_{t+1} \mid C_t).$$

That is, conditioning on the 'history' of the process up to time t is equivalent to conditioning only on the most recent value C_t. For compactness we define $\mathbf{C}^{(t)}$ as the history $(C_1, C_2, \ldots, C_t)$, in which case the Markov property can be written as

$$\Pr(C_{t+1} \mid \mathbf{C}^{(t)}) = \Pr(C_{t+1} \mid C_t).$$

The Markov property can be regarded as a 'first relaxation' of the assumption of independence. The random variables $\{C_t\}$ are dependent in a specific way that is mathematically convenient, as displayed in the following directed graph in which the past and the future are dependent only through the present.

Important quantities associated with a Markov chain are the conditional probabilities called **transition probabilities**:

$$\Pr(C_{s+t} = j \mid C_s = i).$$

If these probabilities do not depend on s, the Markov chain is called **homogeneous**, otherwise nonhomogeneous. Unless there is an explicit indication to the contrary we shall assume that the Markov chain under discussion is homogeneous, in which case the transition probabilities will be denoted by

$$\gamma_{ij}(t) = \Pr(C_{s+t} = j \mid C_s = i).$$

Notice that the notation $\gamma_{ij}(t)$ does not involve s. The matrix $\mathbf{\Gamma}(t)$ is defined as the matrix with (i,j) element $\gamma_{ij}(t)$.

An important property of all finite state-space homogeneous Markov chains is that they satisfy the **Chapman–Kolmogorov equations**:

$$\mathbf{\Gamma}(t + u) = \mathbf{\Gamma}(t)\,\mathbf{\Gamma}(u).$$

The proof requires only the definition of conditional probability and the application of the Markov property: this is Exercise 9. The Chapman–Kolmogorov equations imply that, for all $t \in \mathbb{N}$,

$$\mathbf{\Gamma}(t) = \mathbf{\Gamma}(1)^t;$$

i.e. the matrix of t-step transition probabilities is the tth power of $\mathbf{\Gamma}(1)$, the matrix of one-step transition probabilities. The matrix $\mathbf{\Gamma}(1)$, which will be abbreviated as $\mathbf{\Gamma}$, is a square matrix of probabilities with row sums equal to 1:

$$\mathbf{\Gamma} = \begin{pmatrix} \gamma_{11} & \cdots & \gamma_{1m} \\ \vdots & \ddots & \vdots \\ \gamma_{m1} & \cdots & \gamma_{mm} \end{pmatrix},$$

where (throughout this text) m denotes the number of states of the Markov chain. The statement that the row sums are equal to 1 can be written as $\mathbf{\Gamma}\mathbf{1}' = \mathbf{1}'$; that is, the column vector $\mathbf{1}'$ is a right eigenvector of $\mathbf{\Gamma}$ and corresponds to eigenvalue 1. We shall refer to $\mathbf{\Gamma}$ as the (one-step) **transition probability matrix** (t.p.m.). Many authors use instead the term 'transition matrix'; we avoid that term because of possible confusion with a matrix of transition counts.

The **unconditional probabilities** $\Pr(C_t = j)$ of a Markov chain being in a given state at a given time t are often of interest. We denote these by the row vector

$$\mathbf{u}(t) = (\Pr(C_t = 1), \ldots, \Pr(C_t = m)), \quad t \in \mathbb{N}.$$

We refer to $u(1)$ as the **initial distribution** of the Markov chain. To deduce the distribution at time $t + 1$ from that at t we postmultiply by the transition probability matrix $\mathbf{\Gamma}$:

$$\mathbf{u}(t + 1) = \mathbf{u}(t)\mathbf{\Gamma}. \tag{1.3}$$

The proof of this statement is left as an exercise.

Example: Imagine that the sequence of rainy and sunny days is such that each day's weather depends only on the previous day's, and the transition probabilities are given by the following table.

	day $t + 1$	
day t	rainy	sunny
rainy	0.9	0.1
sunny	0.6	0.4

That is: if today is rainy, the probability that tomorrow will be rainy is 0.9; if today is sunny, that probability is 0.6. The weather is then a two-state homogeneous Markov chain, with t.p.m. $\mathbf{\Gamma}$ given by

$$\mathbf{\Gamma} = \begin{pmatrix} 0.9 & 0.1 \\ 0.6 & 0.4 \end{pmatrix}.$$

Now suppose that today (time 1) is a sunny day. This means that the distribution of today's weather is

$$\mathbf{u}(1) = \big(\Pr(C_1 = 1), \Pr(C_1 = 2)\big) = \big(0, 1\big).$$

The distribution of the weather of tomorrow, the day after tomorrow, and so on, can be calculated by repeatedly postmultiplying $\mathbf{u}(1)$ by $\mathbf{\Gamma}$, the t.p.m.:

$$\begin{aligned}
\mathbf{u}(2) &= \big(\Pr(C_2 = 1), \Pr(C_2 = 2)\big) = \mathbf{u}(1)\mathbf{\Gamma} = (0.6, 0.4), \\
\mathbf{u}(3) &= \big(\Pr(C_3 = 1), \Pr(C_3 = 2)\big) = \mathbf{u}(2)\mathbf{\Gamma} = (0.78, 0.22), \text{ etc.}
\end{aligned}$$

1.3.2 Stationary distributions

A Markov chain with transition probability matrix $\mathbf{\Gamma}$ is said to have **stationary distribution** $\boldsymbol{\delta}$ (a row vector with nonnegative elements) if $\boldsymbol{\delta}\mathbf{\Gamma} = \boldsymbol{\delta}$ and $\boldsymbol{\delta}\mathbf{1}' = 1$. The first of these requirements expresses the stationarity, the second is the requirement that $\boldsymbol{\delta}$ is indeed a probability distribution.

For instance, the Markov chain with t.p.m. given by

$$\mathbf{\Gamma} = \begin{pmatrix} 1/3 & 1/3 & 1/3 \\ 2/3 & 0 & 1/3 \\ 1/2 & 1/2 & 0 \end{pmatrix}$$

has as stationary distribution $\boldsymbol{\delta} = \frac{1}{32}(15, 9, 8)$.

Since $\mathbf{u}(t+1) = \mathbf{u}(t)\mathbf{\Gamma}$, a Markov chain started from its stationary distribution will continue to have that distribution at all subsequent time points, and we shall refer to such a process as a **stationary Markov chain**. It is perhaps worth stating that this assumes more than merely homogeneity; homogeneity alone would not be sufficient to render the Markov chain a stationary process, and we prefer to reserve the adjective 'stationary' for homogeneous Markov chains that have the additional property that the initial distribution $\mathbf{u}(1)$ is the stationary distribution. Not all authors use this terminology, however: see e.g. McLachlan and Peel (2000, p. 328), who use the word 'stationary' of a Markov chain where we would say 'homogeneous'.

An irreducible (homogeneous, discrete-time, finite state-space) Markov chain has a unique, strictly positive, stationary distribution. Note that although the technical assumption of irreducibility is needed for this conclusion, aperiodicity is not: see Grimmett and Stirzaker (2001), Lemma 6.3.5 on p. 225 and Theorem 6.4.3 on p. 227.

A general result that can conveniently be used to compute a stationary distribution (see Exercise 8(a)) is as follows. The vector $\boldsymbol{\delta}$ with nonnega-

tive elements is a stationary distribution of the Markov chain with t.p.m.
$\boldsymbol{\Gamma}$ if and only if

$$\boldsymbol{\delta}(\mathbf{I}_m - \boldsymbol{\Gamma} + \mathbf{U}) = \mathbf{1},$$

where $\mathbf{1}$ is a row vector of ones, $\mathbf{I}_m$ is the $m \times m$ identity matrix, and
$\mathbf{U}$ is the $m \times m$ matrix of ones. Alternatively, a stationary distribution
can be found by deleting one of the equations in the system $\boldsymbol{\delta}\boldsymbol{\Gamma} = \boldsymbol{\delta}$ and
replacing it by $\sum_i \delta_i = 1$.

1.3.3 Reversibility

A property of Markov chains (and other random processes) that is sometimes of interest is reversibility. A random process is said to be **reversible** if its finite-dimensional distributions are invariant under rever-
sal of time. In the case of a stationary irreducible Markov chain with
t.p.m. $\boldsymbol{\Gamma}$ and stationary distribution $\boldsymbol{\delta}$, it is necessary and sufficient for
reversibility that the 'detailed balance conditions'

$$\delta_i \gamma_{ij} = \delta_j \gamma_{ji}$$

be satisfied for all states i and j (Kelly, 1979, p. 5). These conditions are
trivially satisfied by all two-state stationary irreducible Markov chains,
which are thereby reversible. The Markov chain in the example in Section
1.3.2 is not reversible, however, because $\delta_1 \gamma_{12} = \frac{15}{32} \times \frac{1}{3} = \frac{5}{32}$ but $\delta_2 \gamma_{21} = \frac{9}{32} \times \frac{2}{3} = \frac{6}{32}$. Exercise 8 in Chapter 2 and Exercise 8 in Chapter 9 present
some results and examples concerning reversibility in HMMs.

1.3.4 Autocorrelation function

We shall have occasion (e.g. in Sections 10.2.1 and 13.1.1) to compare
the autocorrelation function (ACF) of a hidden Markov model with that
of a Markov chain. We therefore discuss the latter now. We assume of
course that the states of the Markov chain are quantitative and not
merely categorical. The ACF of a Markov chain $\{C_t\}$ on $\{1, 2, \ldots, m\}$,
assumed stationary and irreducible, may be obtained as follows.

Firstly, defining $\mathbf{v} = (1, 2, \ldots, m)$ and $\mathbf{V} = \text{diag}(1, 2, \ldots, m)$, we have,
for all nonnegative integers k,

$$\text{Cov}(C_t, C_{t+k}) = \boldsymbol{\delta}\mathbf{V}\boldsymbol{\Gamma}^k\mathbf{v}' - (\boldsymbol{\delta}\mathbf{v}')^2; \qquad (1.4)$$

the proof is Exercise 10. Secondly, if $\boldsymbol{\Gamma}$ is diagonalizable, and its eigenvalues (other than 1) are denoted by $\omega_2, \omega_3, \ldots, \omega_m$, then $\boldsymbol{\Gamma}$ can be
written as

$$\boldsymbol{\Gamma} = \mathbf{U}\boldsymbol{\Omega}\mathbf{U}^{-1},$$

where $\boldsymbol{\Omega}$ is $\text{diag}(1, \omega_2, \omega_3, \ldots, \omega_m)$ and the columns of $\mathbf{U}$ are corresponding right eigenvectors of $\boldsymbol{\Gamma}$. We then have, for nonnegative integers k,

$$
\begin{aligned}
\mathrm{Cov}(C_t, C_{t+k}) &= \boldsymbol{\delta}\mathbf{V}\mathbf{U}\boldsymbol{\Omega}^k\mathbf{U}^{-1}\mathbf{v}' - (\boldsymbol{\delta}\mathbf{v}')^2 \\
&= \mathbf{a}\boldsymbol{\Omega}^k\mathbf{b}' - a_1 b_1 \\
&= \sum_{i=2}^{m} a_i b_i \omega_i^k,
\end{aligned}
$$

where $\mathbf{a} = \boldsymbol{\delta}\mathbf{V}\mathbf{U}$ and $\mathbf{b}' = \mathbf{U}^{-1}\mathbf{v}'$. Hence $\mathrm{Var}(C_t) = \sum_{i=2}^{m} a_i b_i$ and, for nonnegative integers k,

$$
\rho(k) \equiv \mathrm{Corr}(C_t, C_{t+k}) = \sum_{i=2}^{m} a_i b_i \omega_i^k \Big/ \sum_{i=2}^{m} a_i b_i. \tag{1.5}
$$

This is a weighted average of the kth powers of the eigenvalues ω_2, ω_3, $\ldots$, ω_m, and somewhat similar to the ACF of a Gaussian autoregressive process of order $m - 1$. Note that Equation (1.5) implies in the case $m = 2$ that $\rho(k) = \rho(1)^k$ for all nonnegative integers k, and that $\rho(1)$ is the eigenvalue other than 1 of $\boldsymbol{\Gamma}$.

1.3.5 Estimating transition probabilities

If we are given a realization of a Markov chain, and wish to estimate the transition probabilities, one approach — but not the only one — is to find the transition counts and estimate the transition probabilities from them in an obvious way. For instance, if the MC has three states and the observed sequence is

2332111112 3132332122 3232332222 3132332212 3232132232
3132332223 3232331232 3232331222 3232132123 3132332121,

then the matrix of transition counts is

$$
(f_{ij}) = \begin{pmatrix} 4 & 7 & 6 \\ 8 & 10 & 24 \\ 6 & 24 & 10 \end{pmatrix},
$$

where f_{ij} denotes the number of transitions observed from state i to state j. Since the number of transitions from state 2 to state 3 is 24, and the total number of transitions from state 2 is $8+10+24$, a plausible estimate of γ_{23} is $24/42$. The t.p.m. $\boldsymbol{\Gamma}$ is therefore plausibly estimated by

$$
\begin{pmatrix} 4/17 & 7/17 & 6/17 \\ 8/42 & 10/42 & 24/42 \\ 6/40 & 24/40 & 10/40 \end{pmatrix}.
$$

We shall now show that this is in fact the conditional maximum likelihood estimate of $\boldsymbol{\Gamma}$, conditioned on the first observation.

Suppose, then, that we wish to estimate the $m^2 - m$ parameters γ_{ij} ($i \neq j$) of an m-state Markov chain $\{C_t\}$ from a realization $c_1, c_2, \ldots, c_T$. The likelihood conditioned on the first observation is

$$L = \prod_{i=1}^{m} \prod_{j=1}^{m} \gamma_{ij}^{f_{ij}}.$$

The log-likelihood is

$$l = \sum_{i=1}^{m} \left(\sum_{j=1}^{m} f_{ij} \log \gamma_{ij} \right) = \sum_{i=1}^{m} l_i \text{ (say)},$$

and we can maximize l by maximizing each l_i separately. Substituting $1 - \sum_{k \neq i} \gamma_{ik}$ for γ_{ii}, differentiating l_i with respect to an off-diagonal transition probability γ_{ij}, and equating the derivative to zero yields

$$0 = \frac{-f_{ii}}{1 - \sum_{k \neq i} \gamma_{ik}} + \frac{f_{ij}}{\gamma_{ij}} = -\frac{f_{ii}}{\gamma_{ii}} + \frac{f_{ij}}{\gamma_{ij}}.$$

Hence, unless a denominator is zero in the above equation, $f_{ij}\gamma_{ii} = f_{ii}\gamma_{ij}$, and so $\gamma_{ii} \sum_{j=1}^{m} f_{ij} = f_{ii}$. This implies that, at a (local) maximum of the likelihood,

$$\gamma_{ii} = f_{ii} \Big/ \sum_{j=1}^{m} f_{ij} \quad \text{and} \quad \gamma_{ij} = f_{ij}\gamma_{ii}/f_{ii} = f_{ij} \Big/ \sum_{j=1}^{m} f_{ij}.$$

(We could instead use Lagrange multipliers to express the constraints $\sum_{j=1}^{m} \gamma_{ij} = 1$ subject to which we seek to maximize the terms l_i and therefore the likelihood: see Exercise 11.)

The estimator $\hat{\gamma}_{ij} = f_{ij}/\sum_{k=1}^{m} f_{ik}$ ($i, j = 1, \ldots, m$) — which is just the empirical transition probability — is thereby seen to be a conditional maximum likelihood estimator of γ_{ij}. Note also that this estimator of Γ satisfies the requirement that the row sums should be equal to 1.

The assumption of stationarity of the Markov chain was not used in the above derivation. If we are prepared to assume stationarity, we may use the unconditional likelihood. This is the conditional likelihood as above, multiplied by the stationary probability δ_{c_1}. The unconditional likelihood or its logarithm may then be maximized numerically, subject to nonnegativity and row-sum constraints, in order to estimate the transition probabilities γ_{ij}. Bisgaard and Travis (1991) show in the case of a two-state Markov chain that, barring some extreme cases, the unconditional likelihood equations have a unique solution. For some nontrivial special cases of the two-state chain, they also derive explicit expressions for the unconditional MLEs of the transition probabilities. Since we use this result later (in Section 10.2.1), we state it here.

Suppose the Markov chain $\{C_t\}$ takes the values 0 and 1, and that we wish to estimate the transition probabilities γ_{ij} from a sequence of observations in which there are f_{ij} transitions from state i to state j $(i, j = 0, 1)$, but $f_{00} = 0$. So in the observations a zero is always followed by a one. Define $c = f_{10} + (1 - c_1)$ and $d = f_{11}$. Then the unconditional MLEs of the transition probabilities are given by

$$\hat{\gamma}_{01} = 1 \quad \text{and} \quad \hat{\gamma}_{10} = \frac{-(1 + d) + \left((1 + d)^2 + 4c(c + d - 1)\right)^{\frac{1}{2}}}{2(c + d - 1)}. \quad (1.6)$$

1.3.6 Higher-order Markov chains

This section is somewhat specialized, and the material is used only in Section 8.3 and parts of Sections 10.2.2 and 12.2.2. It will therefore not interrupt the continuity greatly if the reader should initially omit this section.

In cases where observations on a process with finite state-space appear not to satisfy the Markov property, one possibility that suggests itself is to use a higher-order Markov chain, i.e. a model $\{C_t\}$ satisfying the following generalization of the Markov property for some $l \geq 2$:

$$\Pr(C_t \mid C_{t-1}, C_{t-2}, \ldots) = \Pr(C_t \mid C_{t-1}, \ldots, C_{t-l}).$$

An account of such higher-order Markov chains may be found, for instance, in Lloyd (1980, Section 19.9). Although such a model is not in the usual sense a Markov chain, i.e. not a 'first-order' Markov chain, we can redefine the model in such a way as to produce an equivalent process which is. If we let $\mathbf{Y}_t = (C_{t-l+1}, C_{t-l+2}, \ldots, C_t)$, then $\{\mathbf{Y}_t\}$ is a first-order Markov chain on M^l, where M is the state space of $\{C_t\}$. Although some properties may be more awkward to establish, no essentially new theory is therefore involved in analysing a higher-order Markov chain rather than a first-order one.

A *second-order* Markov chain, if stationary, is characterized by the transition probabilities

$$\gamma(i, j, k) = \mathrm{P}(C_t = k \mid C_{t-1} = j, C_{t-2} = i),$$

and has stationary bivariate distribution $u(j, k) = \mathrm{P}(C_{t-1} = j, C_t = k)$ satisfying

$$u(j, k) = \sum_{i=1}^{m} u(i, j)\gamma(i, j, k) \quad \text{and} \quad \sum_{j=1}^{m} \sum_{k=1}^{m} u(j, k) = 1.$$

For example, the most general stationary second-order Markov chain $\{C_t\}$ on the two states 1 and 2 is characterized by the following four transition probabilities:

$$
\begin{aligned}
a &= \Pr(C_t{=}2 \mid C_{t-1}{=}1, C_{t-2}{=}1),\\
b &= \Pr(C_t{=}1 \mid C_{t-1}{=}2, C_{t-2}{=}2),\\
c &= \Pr(C_t{=}1 \mid C_{t-1}{=}2, C_{t-2}{=}1),\\
d &= \Pr(C_t{=}2 \mid C_{t-1}{=}1, C_{t-2}{=}2).
\end{aligned}
$$

The process $\{\mathbf{Y}_t\} = \{(C_{t-1}, C_t)\}$ is then a first-order Markov chain, on the four states (1,1), (1,2), (2,1), (2,2), with transition probability matrix

$$
\begin{pmatrix}
1-a & a & 0 & 0 \\
0 & 0 & c & 1-c \\
1-d & d & 0 & 0 \\
0 & 0 & b & 1-b
\end{pmatrix}. \tag{1.7}
$$

Notice the structural zeros appearing in this matrix. It is not possible, for instance, to make a transition directly from (2,1) to (2,2); hence the zero in row 3 and column 4 in the t.p.m. (1.7). The parameters a, b, c and d are bounded by 0 and 1 but are otherwise unconstrained. The stationary distribution of $\{\mathbf{Y}_t\}$ is proportional to the vector

$$
\bigl(b(1-d), ab, ab, a(1-c)\bigr),
$$

from which it follows that the matrix $(u(j,k))$ of stationary bivariate probabilities for $\{C_t\}$ is

$$
\frac{1}{b(1-d) + 2ab + a(1-c)} \begin{pmatrix} b(1-d) & ab \\ ab & a(1-c) \end{pmatrix}.
$$

Of course the use of a general higher-order Markov chain (instead of a first-order one) increases the number of parameters of the model; a general Markov chain of order l on m states has $m^l(m-1)$ independent transition probabilities. Pegram (1980) and Raftery (1985a,b) have therefore proposed certain classes of parsimonious models for higher-order chains. Pegram's models have $m + l - 1$ parameters, and those of Raftery $m(m-1) + l - 1$. For $m = 2$ the models are equivalent, but for $m > 2$ those of Raftery are more general and can represent a wider range of dependence patterns and autocorrelation structures. In both cases an increase of one in the order of the Markov chain requires only one additional parameter.

Raftery's models, which he terms 'mixture transition distribution' (MTD) models, are defined as follows. The process $\{C_t\}$ takes values in $M = \{1, 2, \ldots, m\}$ and satisfies

$$
\Pr(C_t{=}j_0 \mid C_{t-1}{=}j_1, \ldots, C_{t-l}{=}j_l) = \sum_{i=1}^{l} \lambda_i\, q(j_i, j_0), \tag{1.8}
$$

where $\sum_{i=1}^{l} \lambda_i = 1$, and $\mathbf{Q} = (q(j,k))$ is an $m \times m$ matrix with non-negative entries and row sums equal to one, such that the right-hand side of Equation (1.8) is bounded by zero and one for all $j_0, j_1, \ldots,$ $j_l \in M$. This last requirement, which generates m^{l+1} pairs of nonlinear constraints on the parameters, ensures that the conditional probabilities in Equation (1.8) are indeed probabilities, and the condition on the row sums of $\mathbf{Q}$ ensures that the sum over j_0 of these conditional probabilities is one. Note that Raftery does not assume that the parameters λ_i are nonnegative.

A variety of applications are presented by Raftery (1985a) and by Raftery and Tavaré (1994). In several of the fitted models there are negative estimates of some of the coefficients λ_i. For further accounts of this class of models, see Haney (1993), Berchtold (2001), and Berchtold and Raftery (2002).

Azzalini and Bowman (1990) report the fitting of a second-order Markov chain model to the binary series they use to represent the lengths of successive eruptions of the Old Faithful geyser. Their analysis, and some alternative models, will be discussed in Chapter 10.

Exercises

1. Let X be a random variable which is distributed as a δ_1, δ_2-mixture of two distributions with expectations μ_1, μ_2, and variances σ_1^2 and σ_2^2, respectively, where $\delta_1 + \delta_2 = 1$.

   (a) Show that $\mathrm{Var}(X) = \delta_1 \sigma_1^2 + \delta_2 \sigma_2^2 + \delta_1 \delta_2 (\mu_1 - \mu_2)^2$.

   (b) Show that a (nontrivial) mixture X of two Poisson distributions with distinct means is overdispersed, that is $\mathrm{Var}(X) > \mathrm{E}(X)$.

   (c) Generalize part (b) to a mixture of $m \geq 2$ Poisson distributions.

2. A zero-inflated Poisson distribution is sometimes used as a model for unbounded counts displaying overdispersion relative to the Poisson. Such a model is a mixture of two distributions: one is a Poisson and the other is identically zero.

   (a) Is it ever possible for such a model to display *under*dispersion?

   (b) Now consider the zero-inflated binomial. Is it possible in such a model that the variance is less than the mean?

3. Write an **R** function to minimize minus the log-likelihood of a Poisson mixture model with m components, using the nonlinear minimizer `nlm`.

   Hint: first write a function to transform the parameters $\boldsymbol{\delta}$ and $\boldsymbol{\lambda}$ into the unconstrained parameters $\boldsymbol{\tau}$ and $\boldsymbol{\eta}$ defined on p. 10. You will also need a function to reverse the transformation.

4. Consider the following data, which appear in Lange (1995, 2002, 2004). (There they are quoted from Titterington, Smith and Makov (1985) and Hasselblad (1969), but the trail leads back via Schilling (1947) to Whitaker (1914), where in all eight similar datasets appear as Table XV on p. 67.)

Here n_i denotes the number of days in 1910–1912 on which there appeared, in *The Times* of London, i death notices in respect of women aged 80 or over at death.

i	0	1	2	3	4	5	6	7	8	9
n_i	162	267	271	185	111	61	27	8	3	1

(a) Use `nlm` or `optim` in **R** to fit a mixture of two Poisson distributions to these observations. (The parameter estimates reported by Lange (2002, p. 36; 2004, p. 151) are, in our notation: $\hat{\delta}_1 = 0.3599$, $\hat{\lambda}_1 = 1.2561$ and $\hat{\lambda}_2 = 2.6634$.)

(b) Fit also a single Poisson distribution to these data. Is a single Poisson distribution adequate as a model?

(c) Fit a mixture of three Poisson distributions to these observations.

(d) How many components do you think are necessary?

(e) Repeat (a)–(d) for some of the other seven datasets of Whitaker.

5. Consider the series of weekly sales (in integer units) of a particular soap product in a supermarket, as shown in Table 1.5. The data are taken from a database provided by the Kilts Center for Marketing, Graduate School of Business of the University of Chicago, at: `http://gsbwww.uchicago.edu/kilts/research/db/dominicks`. (The product is 'Zest White Water 15 oz.', with code 3700031165.)

Fit Poisson mixture models with one, two, three and four components. How many components do you think are necessary?

6. Consider a stationary two-state Markov chain with transition probability matrix given by

$$\mathbf{\Gamma} = \begin{pmatrix} \gamma_{11} & \gamma_{12} \\ \gamma_{21} & \gamma_{22} \end{pmatrix}.$$

(a) Show that the stationary distribution is

$$(\delta_1, \delta_2) = \frac{1}{\gamma_{12} + \gamma_{21}} (\gamma_{21}, \gamma_{12}) .$$

(b) Consider the case

$$\mathbf{\Gamma} = \begin{pmatrix} 0.9 & 0.1 \\ 0.2 & 0.8 \end{pmatrix},$$

Table 1.5 *Weekly sales of the soap product; to be read across rows.*

1	6	9	18	14	8	8	1	6	7	3	3	1	3	4	12	8	10	8	2	
17	15	7	12	22	10	4	7	5	0	2	5	3	4	4	7	5	6	1	3	
4	5	3	7	3	0	4	5	3	3	4	4	4	4	4	3	5	5	5	7	
4	0	4	3	2	6	3	8	9	6	3	4	3	3	3	3	2	1	4	5	
5	2	7	5	2	3	1	3	4	6	8	8	5	7	2	4	2	7	4	15	
15	12	21	20	13	9	8	0	13	9	8	0	6	2	0	3	2	4	4	6	
3	2	5	5	3	2	1	1	3	1	2	6	2	7	3	2	4	1	5	6	
8	14	5	3	6	5	11	4	5	9	9	7	9	8	3	4	8	6	3	5	
6	3	1	7	4	9	2	6	6	4	6	6	13	7	4	8	6	4	4	4	
9	2	9	2	2	2	13	13	4	5	1	4	6	5	4	2	3	10	6	15	
5	9	9	7	4	4	2	4	2	3	8	15	0	0	3	4	3	4	7	5	
7	6	0	6	4	14	5	1	6	5	5	4	9	4	14	2	2	1	5	2	
6	4																			

and the following two sequences of observations that are assumed to be generated by the above Markov chain.

$$\text{Sequence 1:} \quad 1 \quad 1 \quad 1 \quad 2 \quad 2 \quad 1$$
$$\text{Sequence 2:} \quad 2 \quad 1 \quad 1 \quad 2 \quad 1 \quad 1$$

Compute the probability of each of the sequences. Note that each sequence contains the same number of ones and twos. Why are these sequences not equally probable?

7. Consider a stationary two-state Markov chain with transition probability matrix given by

$$\mathbf{\Gamma} = \begin{pmatrix} \gamma_{11} & \gamma_{12} \\ \gamma_{21} & \gamma_{22} \end{pmatrix}.$$

Show that the k-step transition probability matrix, i.e. $\mathbf{\Gamma}^k$, is given by

$$\mathbf{\Gamma}^k = \begin{pmatrix} \delta_1 & \delta_2 \\ \delta_1 & \delta_2 \end{pmatrix} + w^k \begin{pmatrix} \delta_2 & -\delta_2 \\ -\delta_1 & \delta_1 \end{pmatrix},$$

where $w = 1 - \gamma_{12} - \gamma_{21}$ and δ_1 and δ_2 are as defined in Exercise 6. (Hint: One way of showing this is to diagonalize the transition probability matrix. But there is a quicker way.)

8.(a) This is one of several possible approaches to finding the stationary distribution of a Markov chain; plundered from Grimmett and Stirzaker (2001), Exercise 6.6.5.

Suppose $\mathbf{\Gamma}$ is the transition probability matrix of a (discrete-time, homogeneous) Markov chain on m states, and that δ is a nonnegative row vector with m components. Show that δ is a stationary distribution of the Markov chain if and only if

$$\delta(\mathbf{I}_m - \mathbf{\Gamma} + \mathbf{U}) = \mathbf{1},$$

where $\mathbf{1}$ is a row vector of ones, and $\mathbf{U}$ is an $m \times m$ matrix of ones.

(b) Write an **R** function `statdist(gamma)` that computes the stationary distribution of the Markov chain with t.p.m. `gamma`.

(c) Use your function to find stationary distributions corresponding to the following transition probability matrices. One of them should cause a problem!

i.
$$\begin{pmatrix} 0.7 & 0.2 & 0.1 \\ 0 & 0.6 & 0.4 \\ 0.5 & 0 & 0.5 \end{pmatrix}$$

ii.
$$\begin{pmatrix} 0 & 1 & 0 \\ \frac{1}{3} & 0 & \frac{2}{3} \\ 0 & 1 & 0 \end{pmatrix}$$

iii.
$$\begin{pmatrix} 0 & 0.5 & 0 & 0.5 \\ 0.75 & 0 & 0.25 & 0 \\ 0 & 0.75 & 0 & 0.25 \\ 0.5 & 0 & 0.5 & 0 \end{pmatrix}$$

iv.
$$\begin{pmatrix} 0.25 & 0.25 & 0.25 & 0.25 \\ 0.25 & 0.25 & 0.5 & 0 \\ 0 & 0 & 0.25 & 0.75 \\ 0 & 0 & 0.5 & 0.5 \end{pmatrix}$$

v.
$$\begin{pmatrix} 1 & 0 & 0 & 0 \\ 0.5 & 0 & 0.5 & 0 \\ 0 & 0.75 & 0 & 0.25 \\ 0 & 0 & 0 & 1 \end{pmatrix}$$

9. Prove the Chapman–Kolmogorov equations.

10. Prove Equation (1.4).

11. Let the quantities a_i be nonnegative, with $\sum_i a_i > 0$. Using a Lagrange multiplier, maximize $S = \sum_{i=1}^m a_i \log \delta_i$ over $\delta_i \geq 0$, subject to $\sum_i \delta_i = 1$. (Check the second- as well as the first-derivative condition.)

12. (This is based on Example 2 of Bisgaard and Travis (1991).) Consider the following sequence of 21 observations, assumed to arise from a two-state (homogeneous) Markov chain:

$$11101\ 10111\ 10110\ 11111\ 1.$$

(a) Estimate the transition probability matrix by ML (maximum likelihood) conditional on the first observation.

(b) Estimate the t.p.m. by unconditional ML (assuming stationarity of the Markov chain).

(c) Use the **R** functions contour and persp to produce contour and perspective plots of the unconditional log-likelihood (as a function of the two off-diagonal transition probabilities).

13. Consider the following two transition probability matrices, neither of which is diagonalizable:

(a)

$$\Gamma = \begin{pmatrix} 1/3 & 1/3 & 1/3 \\ 2/3 & 0 & 1/3 \\ 1/2 & 1/2 & 0 \end{pmatrix};$$

(b)

$$\Gamma = \begin{pmatrix} 0.9 & 0.08 & 0 & 0.02 \\ 0 & 0.7 & 0.2 & 0.1 \\ 0 & 0 & 0.7 & 0.3 \\ 0 & 0 & 0 & 1 \end{pmatrix}.$$

In each case, write Γ in Jordan canonical form, and so find an explicit expression for the t-step transition probabilities ($t = 1, 2, \dots$).

14. Consider the following (very) short DNA sequence, taken from Singh (2003, p. 358):

AACGT CTCTA TCATG CCAGG ATCTG

Fit a homogeneous Markov chain to these data by

(a) maximizing the likelihood conditioned on the first observation;

(b) assuming stationarity and maximizing the unconditional likelihood of all 25 observations.

Compare your estimates of the t.p.m. with each other and with the estimate displayed in Table 1 of Singh (p. 360).

15. Write an **R** function rMC(n,m,gamma,delta=NULL) that generates a series of length n from an m-state Markov chain with t.p.m. gamma. If the initial state distribution is given, then it should be used; otherwise the stationary distribution should be used as the initial distribution. (Use your function statdist from Exercise 8(b).)

CHAPTER 2

Hidden Markov models: definition and properties

2.1 A simple hidden Markov model

Consider again the observed earthquake series displayed in Figure 1.1 on p. 4. The observations are unbounded counts, making the Poisson distribution a natural choice to describe them. However, the sample variance of the observations is substantially greater than the sample mean, indicating overdispersion relative to the Poisson. In Exercise 1 of Chapter 1 we saw that one can accommodate overdispersion by using a mixture model, specifically a mixture of Poisson distributions for this series.

We suppose that each count is generated by one of m Poisson distributions, with means $\lambda_1, \lambda_2, \ldots, \lambda_m$, where the choice of mean is made by a second random mechanism, the parameter process. The mean λ_i is selected with probability δ_i, where $i = 1, 2, \ldots, m$ and $\sum_{i=1}^{m} \delta_i = 1$. The variance of the mixture model is greater than its expectation, which takes care of the problem of overdispersion.

An independent mixture model will not do for the earthquake series because — by definition — it does not allow for the serial dependence in the observations. The sample autocorrelation function, displayed in

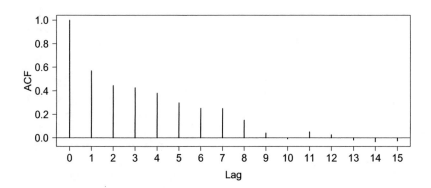

Figure 2.1 *Earthquakes series: sample autocorrelation function (ACF).*

Figure 2.1, gives a clear indication that the observations are serially dependent. One way of allowing for serial dependence in the observations is to relax the assumption that the parameter process is serially independent. A simple and mathematically convenient way to do so is to assume that it is a Markov chain. The resulting model for the observations is called a Poisson–hidden Markov model, a simple example of the class of models discussed in the rest of this book, namely hidden Markov models (HMMs).

We shall not give an account here of the (interesting) history of such models, but two valuable sources of information on HMMs that go far beyond the scope of this book, and include accounts of the history, are Ephraim and Merhav (2002) and Cappé, Moulines and Rydén (2005).

2.2 The basics

2.2.1 Definition and notation

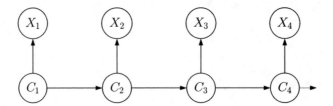

Figure 2.2 *Directed graph of basic HMM.*

A **hidden Markov model** $\{X_t : t \in \mathbb{N}\}$ is a particular kind of dependent mixture. With $\mathbf{X}^{(t)}$ and $\mathbf{C}^{(t)}$ representing the histories from time 1 to time t, one can summarize the simplest model of this kind by:

$$\Pr(C_t \mid \mathbf{C}^{(t-1)}) = \Pr(C_t \mid C_{t-1}), \; t = 2, 3, \ldots \qquad (2.1)$$
$$\Pr(X_t \mid \mathbf{X}^{(t-1)}, \mathbf{C}^{(t)}) = \Pr(X_t \mid C_t), \; t \in \mathbb{N}. \qquad (2.2)$$

The model consists of two parts: firstly, an unobserved 'parameter process' $\{C_t : t = 1, 2, \ldots \}$ satisfying the Markov property, and secondly the 'state-dependent process' $\{X_t : t = 1, 2, \ldots \}$ such that, when C_t is known, the distribution of X_t depends only on the current state C_t and not on previous states or observations. This structure is represented by the directed graph in Figure 2.2. If the Markov chain $\{C_t\}$ has m states, we call $\{X_t\}$ an m-state HMM. Although it is the usual terminology in speech-processing applications, the name 'hidden Markov model' is by no means the only one used for such models or similar ones. For instance, Ephraim and Merhav (2002) argue for 'hidden Markov process', Leroux and Puterman (1992) use 'Markov-dependent mixture', and others use 'Markov-switching model' (especially for models with extra dependencies at the level of the observations X_t), 'models subject to Markov regime', or 'Markov mixture model'.

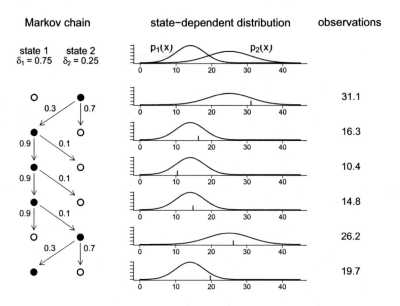

Figure 2.3 *Process generating the observations in a two-state HMM. The chain followed the path 2,1,1,1,2,1, as indicated on the left. The corresponding state-dependent distributions are shown in the middle. The observations are generated from the corresponding active distributions.*

The process generating the observations is demonstrated again in Figure 2.3, for state-dependent distributions p_1 and p_2, stationary distribution $\boldsymbol{\delta} = (0.75, 0.25)$, and t.p.m. $\boldsymbol{\Gamma} = \begin{pmatrix} 0.9 & 0.1 \\ 0.3 & 0.7 \end{pmatrix}$. In contrast to the case of an independent mixture, here the distribution of C_t, the state at time t, does depend on C_{t-1}. As is also true of independent mixtures, there is for each state a different distribution, discrete or continuous.

We now introduce some notation which will cover both discrete- and continuous-valued observations. In the case of discrete observations we define, for $i = 1, 2, \ldots, m$,

$$p_i(x) = \Pr(X_t = x \mid C_t = i).$$

That is, p_i is the probability mass function of X_t if the Markov chain is in state i at time t. The continuous case is treated similarly: there we define p_i to be the probability *density* function of X_t if the Markov chain is in state i at time t. We refer to the m distributions p_i as the **state-dependent distributions** of the model. Many of our results are stated only in the discrete form, but, if probabilities are interpreted as densities, apply also to the continuous case.

2.2.2 Marginal distributions

We shall often need the distribution of X_t and also higher-order marginal distributions, such as that of (X_t, X_{t+k}). We shall derive the results for the case in which the Markov chain is homogeneous but not necessarily stationary, and then give them as well for the special case in which the Markov chain is stationary. For convenience the derivation is given only for discrete state-dependent distributions; the continuous case can be derived analogously.

Univariate distributions

For discrete-valued observations X_t, defining $u_i(t) = \Pr(C_t = i)$ for $t = 1, \ldots, T$, we have

$$\Pr(X_t = x) = \sum_{i=1}^{m} \Pr(C_t = i)\Pr(X_t = x \mid C_t = i)$$

$$= \sum_{i=1}^{m} u_i(t)p_i(x).$$

This expression can conveniently be rewritten in matrix notation:

$$\Pr(X_t = x) = (u_1(t), \ldots, u_m(t)) \begin{pmatrix} p_1(x) & & 0 \\ & \ddots & \\ 0 & & p_m(x) \end{pmatrix} \begin{pmatrix} 1 \\ \vdots \\ 1 \end{pmatrix}$$

$$= \mathbf{u}(t)\mathbf{P}(x)\mathbf{1}',$$

where $\mathbf{P}(x)$ is defined as the diagonal matrix with i th diagonal element $p_i(x)$. It follows from Equation (1.3) that $\mathbf{u}(t) = \mathbf{u}(1)\mathbf{\Gamma}^{t-1}$, and hence that

$$\Pr(X_t = x) = \mathbf{u}(1)\mathbf{\Gamma}^{t-1}\mathbf{P}(x)\mathbf{1}'. \tag{2.3}$$

Equation (2.3) holds if the Markov chain is merely homogeneous, and not necessarily stationary. If, as we shall often assume, the Markov chain is stationary, with stationary distribution $\boldsymbol{\delta}$, then the result is simpler: in that case $\boldsymbol{\delta}\mathbf{\Gamma}^{t-1} = \boldsymbol{\delta}$ for all $t \in \mathbb{N}$, and so

$$\Pr(X_t = x) = \boldsymbol{\delta}\mathbf{P}(x)\mathbf{1}'. \tag{2.4}$$

Bivariate distributions

The calculation of many of the distributions relating to an HMM is most easily done by first noting that, in any directed graphical model, the joint distribution of a set of random variables V_i is given by

$$\Pr(V_1, V_2, \ldots, V_n) = \prod_{i=1}^{n} \Pr(V_i \mid \mathrm{pa}(V_i)), \qquad (2.5)$$

where $\mathrm{pa}(V_i)$ denotes all the 'parents' of V_i in the set $V_1, V_2, \ldots, V_n$; see e.g. Davison (2003, p. 250) or Jordan (2004).

Examining the directed graph of the four random variables X_t, X_{t+k}, C_t, C_{t+k}, for positive integer k, we see that $\mathrm{pa}(C_t)$ is empty, $\mathrm{pa}(X_t) = \{C_t\}$, $\mathrm{pa}(C_{t+k}) = \{C_t\}$ and $\mathrm{pa}(X_{t+k}) = \{C_{t+k}\}$. It therefore follows that

$$\Pr(X_t, X_{t+k}, C_t, C_{t+k}) = \Pr(C_t)\Pr(X_t|C_t)\Pr(C_{t+k}|C_t)\Pr(X_{t+k}|C_{t+k}),$$

and hence that

$$
\begin{aligned}
\Pr(X_t = v, & X_{t+k} = w) \\
&= \sum_{i=1}^{m}\sum_{j=1}^{m} \Pr(X_t = v, X_{t+k} = w, C_t = i, C_{t+k} = j) \\
&= \sum_{i=1}^{m}\sum_{j=1}^{m} \underbrace{\Pr(C_t = i)}_{u_i(t)}\, p_i(v)\, \underbrace{\Pr(C_{t+k} = j \mid C_t = i)}_{\gamma_{ij}(k)}\, p_j(w) \\
&= \sum_{i=1}^{m}\sum_{j=1}^{m} u_i(t)p_i(v)\gamma_{ij}(k)p_j(w).
\end{aligned}
$$

Writing the above double sum as a product of matrices yields

$$\Pr(X_t = v, X_{t+k} = w) \;=\; \mathbf{u}(t)\mathbf{P}(v)\mathbf{\Gamma}^k\mathbf{P}(w)\mathbf{1}'. \qquad (2.6)$$

If the Markov chain is stationary, this reduces to

$$\Pr(X_t = v, X_{t+k} = w) \;=\; \boldsymbol{\delta}\mathbf{P}(v)\mathbf{\Gamma}^k\mathbf{P}(w)\mathbf{1}'. \qquad (2.7)$$

Similarly one can obtain expressions for the higher-order marginal distributions; in the stationary case, the formula for a trivariate distribution is, for positive integers k and l,

$$\Pr(X_t = v, X_{t+k} = w, X_{t+k+l} = z) = \boldsymbol{\delta}\mathbf{P}(v)\mathbf{\Gamma}^k\mathbf{P}(w)\mathbf{\Gamma}^l\mathbf{P}(z)\mathbf{1}'.$$

2.2.3 Moments

First we note that

$$E(X_t) = \sum_{i=1}^{m} E(X_t \mid C_t = i) \Pr(C_t = i) = \sum_{i=1}^{m} u_i(t)E(X_t \mid C_t = i),$$

which, in the stationary case, reduces to

$$E(X_t) = \sum_{i=1}^{m} \delta_i E(X_t \mid C_t = i).$$

More generally, analogous results hold for $E(g(X_t))$ and $E(g(X_t, X_{t+k}))$, for any functions g for which the relevant state-dependent expectations exist. In the stationary case

$$E(g(X_t)) = \sum_{i=1}^{m} \delta_i E(g(X_t) \mid C_t = i); \qquad (2.8)$$

and

$$E(g(X_t, X_{t+k})) = \sum_{i,j=1}^{m} E(g(X_t, X_{t+k}) \mid C_t = i, C_{t+k} = j) \, \delta_i \gamma_{ij}(k), \quad (2.9)$$

where $\gamma_{ij}(k) = (\mathbf{\Gamma}^k)_{ij}$, for $k \in \mathbb{N}$. Often we shall be interested in a function g which factorizes as $g(X_t, X_{t+k}) = g_1(X_t)g_2(X_{t+k})$, in which case Equation (2.9) becomes

$$E(g(X_t, X_{t+k})) = \sum_{i,j=1}^{m} E(g_1(X_t) \mid C_t = i) \, E(g_2(X_{t+k}) \mid C_{t+k} = j) \, \delta_i \gamma_{ij}(k).$$
$$(2.10)$$

These expressions enable us, for instance, to find covariances and correlations without too much trouble; convenient explicit expressions exist in many cases. For instance, the following conclusions result in the case of a stationary two-state Poisson–HMM:

- $E(X_t) = \delta_1 \lambda_1 + \delta_2 \lambda_2$;

- $\mathrm{Var}(X_t) = E(X_t) + \delta_1 \delta_2 (\lambda_2 - \lambda_1)^2 \geq E(X_t)$;

- $\mathrm{Cov}(X_t, X_{t+k}) = \delta_1 \delta_2 (\lambda_2 - \lambda_1)^2 (1 - \gamma_{12} - \gamma_{21})^k$, for $k \in \mathbb{N}$.

Notice that the resulting formula for the correlation of X_t and X_{t+k} is of the form $\rho(k) = A(1 - \gamma_{12} - \gamma_{21})^k$ with $A \in [0, 1)$, and that $A = 0$ if $\lambda_1 = \lambda_2$. For more details, and for more general results, see Exercises 3 and 4.

2.3 The likelihood

The aim of this section is to develop an explicit (and computable) formula for the likelihood L_T of T consecutive observations x_1, x_2, ..., x_T assumed to be generated by an m-state HMM. That such a formula exists is indeed fortunate, but by no means obvious. We shall see that the computation of the likelihood, consisting as it does of a sum of m^T terms, each of which is a product of $2T$ factors, appears to require $O(Tm^T)$ operations, and several authors have come to the conclusion that straightforward calculation of the likelihood is infeasible. However, it has long been known in several contexts that the likelihood is computable: see e.g. Baum (1972), Lange and Boehnke (1983), and Cosslett and Lee (1985). What we describe here is in fact a special case of a much more general theory: see Smyth, Heckerman and Jordan (1997) or Jordan (2004).

It is our purpose here to demonstrate that L_T can in general be computed relatively simply in $O(Tm^2)$ operations. Once it is clear that the likelihood is simple to compute, the way will be open to estimate parameters by numerical maximization of the likelihood.

First the likelihood of a two-state model will be explored, and then the general formula will be presented.

2.3.1 The likelihood of a two-state Bernoulli–HMM

Example: Consider the two-state HMM with t.p.m.

$$\mathbf{\Gamma} = \begin{pmatrix} \frac{1}{2} & \frac{1}{2} \\ \frac{1}{4} & \frac{3}{4} \end{pmatrix}$$

and state-dependent distributions given by

$$\Pr(X_t = x \mid C_t = 1) = \frac{1}{2} \quad \text{(for } x = 0, 1)$$

and

$$\Pr(X_t = 1 \mid C_t = 2) = 1.$$

We call a model of this kind a Bernoulli–HMM. The stationary distribution of the Markov chain is $\boldsymbol{\delta} = \frac{1}{3}(1, 2)$. Then the probability that $X_1 = X_2 = X_3 = 1$ can be calculated as follows. First, note that, by Equation (2.5),

$$\Pr(X_1, X_2, X_3, C_1, C_2, C_3)$$
$$= \Pr(C_1) \Pr(X_1 \mid C_1) \Pr(C_2 \mid C_1) \Pr(X_2 \mid C_2) \Pr(C_3 \mid C_2) \Pr(X_3 \mid C_3);$$

Table 2.1 *Example of a likelihood computation.*

i	j	k	$p_i(1)$	$p_j(1)$	$p_k(1)$	δ_i	γ_{ij}	γ_{jk}	product
1	1	1	$\frac{1}{2}$	$\frac{1}{2}$	$\frac{1}{2}$	$\frac{1}{3}$	$\frac{2}{4}$	$\frac{2}{4}$	$\frac{1}{96}$
1	1	2	$\frac{1}{2}$	$\frac{1}{2}$	1	$\frac{1}{3}$	$\frac{2}{4}$	$\frac{2}{4}$	$\frac{1}{48}$
1	2	1	$\frac{1}{2}$	1	$\frac{1}{2}$	$\frac{1}{3}$	$\frac{2}{4}$	$\frac{1}{4}$	$\frac{1}{96}$
1	2	2	$\frac{1}{2}$	1	1	$\frac{1}{3}$	$\frac{2}{4}$	$\frac{3}{4}$	$\frac{1}{16}$
2	1	1	1	$\frac{1}{2}$	$\frac{1}{2}$	$\frac{2}{3}$	$\frac{1}{4}$	$\frac{2}{4}$	$\frac{1}{48}$
2	1	2	1	$\frac{1}{2}$	1	$\frac{2}{3}$	$\frac{1}{4}$	$\frac{2}{4}$	$\frac{1}{24}$
2	2	1	1	1	$\frac{1}{2}$	$\frac{2}{3}$	$\frac{3}{4}$	$\frac{1}{4}$	$\frac{1}{16}$
2	2	2	1	1	1	$\frac{2}{3}$	$\frac{3}{4}$	$\frac{3}{4}$	$\frac{3}{8}$
									$\frac{29}{48}$

and then sum over the values assumed by C_1, C_2, C_3. The result is

$$\Pr(X_1 = 1, X_2 = 1, X_3 = 1)$$
$$= \sum_{i=1}^{2}\sum_{j=1}^{2}\sum_{k=1}^{2} \Pr(X_1 = 1, X_2 = 1, X_3 = 1, C_1 = i, C_2 = j, C_3 = k)$$
$$= \sum_{i=1}^{2}\sum_{j=1}^{2}\sum_{k=1}^{2} \delta_i p_i(1)\gamma_{ij}p_j(1)\gamma_{jk}p_k(1). \qquad (2.11)$$

Notice that the triple sum (2.11) has $m^T = 2^3$ terms, each of which is a product of $2T = 2 \times 3$ factors. To evaluate the required probability, the different possibilities for the values of i, j and k can be listed and the sum (2.11) calculated as in Table 2.1.

Summation of the last column of Table 2.1 tells us that $\Pr(X_1 = 1, X_2 = 1, X_3 = 1) = \frac{29}{48}$. In passing we note that the largest element in the last column is $\frac{3}{8}$; the state sequence ijk that maximizes the joint probability

$$\Pr(X_1 = 1, X_2 = 1, X_3 = 1, C_1 = i, C_2 = j, C_3 = k)$$

is therefore the sequence 222. Equivalently, it maximizes the conditional probability $\Pr(C_1 = i, C_2 = j, C_3 = k \mid X_1 = 1, X_2 = 1, X_3 = 1)$. This is an example of 'global decoding', which will be discussed in Section 5.3.2: see p. 82.

But a more convenient way to present the sum is to use matrix notation. Let $\mathbf{P}(u)$ be defined (as before) as $\mathrm{diag}(p_1(u), p_2(u))$. Then

$$\mathbf{P}(0) = \begin{pmatrix} \frac{1}{2} & 0 \\ 0 & 0 \end{pmatrix} \quad \text{and} \quad \mathbf{P}(1) = \begin{pmatrix} \frac{1}{2} & 0 \\ 0 & 1 \end{pmatrix},$$

and the triple sum (2.11) can be written as

$$\sum_{i=1}^{2} \sum_{j=1}^{2} \sum_{k=1}^{2} \delta_i p_i(1) \gamma_{ij} p_j(1) \gamma_{jk} p_k(1) = \boldsymbol{\delta} \mathbf{P}(1) \boldsymbol{\Gamma} \mathbf{P}(1) \boldsymbol{\Gamma} \mathbf{P}(1) \mathbf{1}'.$$

2.3.2 The likelihood in general

Here we consider the likelihood of an HMM in general. We suppose there is an observation sequence $x_1, x_2, \ldots, x_T$ generated by such a model. We seek the probability L_T of observing that sequence, as calculated under an m-state HMM which has *initial* distribution $\boldsymbol{\delta}$ and t.p.m. $\boldsymbol{\Gamma}$ for the Markov chain, and state-dependent probability (density) functions p_i. In many of our applications we shall assume that $\boldsymbol{\delta}$ is the stationary distribution implied by $\boldsymbol{\Gamma}$, but it is not necessary to make that assumption in general.

Proposition 1 *The likelihood is given by*

$$L_T = \boldsymbol{\delta} \mathbf{P}(x_1) \boldsymbol{\Gamma} \mathbf{P}(x_2) \boldsymbol{\Gamma} \mathbf{P}(x_3) \cdots \boldsymbol{\Gamma} \mathbf{P}(x_T) \mathbf{1}'. \qquad (2.12)$$

If $\boldsymbol{\delta}$, the distribution of C_1, is the stationary distribution of the Markov chain, then in addition

$$L_T = \boldsymbol{\delta} \boldsymbol{\Gamma} \mathbf{P}(x_1) \boldsymbol{\Gamma} \mathbf{P}(x_2) \boldsymbol{\Gamma} \mathbf{P}(x_3) \cdots \boldsymbol{\Gamma} \mathbf{P}(x_T) \mathbf{1}'. \qquad (2.13)$$

Before proving the above proposition, we rewrite the conclusions in a notation which is sometimes useful. For $t = 1, \ldots, T$, let the matrix $\mathbf{B}_t$ be defined by $\mathbf{B}_t = \boldsymbol{\Gamma} \mathbf{P}(x_t)$. Equations (2.12) and (2.13) (respectively) can then be written as

$$L_T = \boldsymbol{\delta} \mathbf{P}(x_1) \mathbf{B}_2 \mathbf{B}_3 \cdots \mathbf{B}_T \mathbf{1}'$$

and

$$L_T = \boldsymbol{\delta} \mathbf{B}_1 \mathbf{B}_2 \mathbf{B}_3 \cdots \mathbf{B}_T \mathbf{1}'.$$

Note that in the first of these equations $\boldsymbol{\delta}$ represents the initial distribution of the Markov chain, and in the second the stationary distribution.

Proof. We present only the case of discrete observations. First note that

$$L_T = \Pr(\mathbf{X}^{(T)} = \mathbf{x}^{(T)}) = \sum_{c_1, c_2, \ldots, c_T = 1}^{m} \Pr(\mathbf{X}^{(T)} = \mathbf{x}^{(T)}, \mathbf{C}^{(T)} = \mathbf{c}^{(T)}),$$

and that, by Equation (2.5),

$$\Pr(\mathbf{X}^{(T)}, \mathbf{C}^{(T)}) = \Pr(C_1) \prod_{k=2}^{T} \Pr(C_k \mid C_{k-1}) \prod_{k=1}^{T} \Pr(X_k \mid C_k). \quad (2.14)$$

It follows that

$$
\begin{aligned}
L_T &= \sum_{c_1,\ldots,c_T=1}^{m} \left(\delta_{c_1} \gamma_{c_1,c_2} \gamma_{c_2,c_3} \cdots \gamma_{c_{T-1},c_T} \right) \left(p_{c_1}(x_1) p_{c_2}(x_2) \cdots p_{c_T}(x_T) \right) \\
&= \sum_{c_1,\ldots,c_T=1}^{m} \delta_{c_1} p_{c_1}(x_1) \gamma_{c_1,c_2} p_{c_2}(x_2) \gamma_{c_2,c_3} \cdots \gamma_{c_{T-1},c_T} p_{c_T}(x_T) \\
&= \boldsymbol{\delta} \mathbf{P}(x_1) \boldsymbol{\Gamma} \mathbf{P}(x_2) \boldsymbol{\Gamma} \mathbf{P}(x_3) \cdots \boldsymbol{\Gamma} \mathbf{P}(x_T) \mathbf{1}',
\end{aligned}
$$

i.e. Equation (2.12). If $\boldsymbol{\delta}$ is the stationary distribution of the Markov chain, we have $\boldsymbol{\delta}\mathbf{P}(x_1) = \boldsymbol{\delta}\boldsymbol{\Gamma}\mathbf{P}(x_1) = \boldsymbol{\delta}\mathbf{B}_1$, hence Equation (2.13), which involves an extra factor of $\boldsymbol{\Gamma}$ but may be slightly simpler to code. □

In order to set out the likelihood computation in the form of an algorithm, let us now define the vector $\boldsymbol{\alpha}_t$, for $t = 1, 2, \ldots, T$, by

$$\boldsymbol{\alpha}_t = \boldsymbol{\delta}\mathbf{P}(x_1)\boldsymbol{\Gamma}\mathbf{P}(x_2)\boldsymbol{\Gamma}\mathbf{P}(x_3) \cdots \boldsymbol{\Gamma}\mathbf{P}(x_t) = \boldsymbol{\delta}\mathbf{P}(x_1) \prod_{s=2}^{t} \boldsymbol{\Gamma}\mathbf{P}(x_s), \quad (2.15)$$

with the convention that an empty product is the identity matrix. It follows immediately from this definition that

$$L_T = \boldsymbol{\alpha}_T \mathbf{1}', \quad \text{and} \quad \boldsymbol{\alpha}_t = \boldsymbol{\alpha}_{t-1}\boldsymbol{\Gamma}\mathbf{P}(x_t) \quad \text{for } t \geq 2.$$

Accordingly, we can conveniently set out as follows the computations involved in the likelihood formula (2.12):

$$\boldsymbol{\alpha}_1 = \boldsymbol{\delta}\mathbf{P}(x_1);$$
$$\boldsymbol{\alpha}_t = \boldsymbol{\alpha}_{t-1}\boldsymbol{\Gamma}\mathbf{P}(x_t) \quad \text{for } t = 2, 3, \ldots, T;$$
$$L_T = \boldsymbol{\alpha}_T \mathbf{1}'.$$

That the number of operations involved is of order Tm^2 can be deduced thus. For each of the values of t in the loop, there are m elements of $\boldsymbol{\alpha}_t$ to be computed, and each of those elements is a sum of m products of three quantities: an element of $\boldsymbol{\alpha}_{t-1}$, a transition probability γ_{ij}, and a state-dependent probability (or density) $p_j(x_t)$.

The corresponding scheme for computation of (2.13) (i.e. if $\boldsymbol{\delta}$, the distribution of C_1, is the stationary distribution of the Markov chain) is

$$\boldsymbol{\alpha}_0 = \boldsymbol{\delta};$$
$$\boldsymbol{\alpha}_t = \boldsymbol{\alpha}_{t-1}\boldsymbol{\Gamma}\mathbf{P}(x_t) \quad \text{for } t = 1, 2, \ldots, T;$$
$$L_T = \boldsymbol{\alpha}_T \mathbf{1}'.$$

The elements of $\boldsymbol{\alpha}_t$ are usually referred to as **forward probabilities**; the reason for this name will appear only later, in Section 4.1.1.

HMMs are not Markov processes

HMMs do not in general satisfy the Markov property. This we can now establish via a simple counterexample. Let X_t and C_t be as defined in the example in Section 2.3.1. We already know that

$$\Pr(X_1 = 1, X_2 = 1, X_3 = 1) = \frac{29}{48},$$

and from the above general expression for the likelihood, or otherwise, it can be established that $\Pr(X_2 = 1) = \frac{5}{6}$, and that

$$\Pr(X_1 = 1, X_2 = 1) = \Pr(X_2 = 1, X_3 = 1) = \frac{17}{24}.$$

It therefore follows that

$$
\begin{aligned}
\Pr(X_3 = 1 \mid X_1 = 1, X_2 = 1) &= \frac{\Pr(X_1 = 1, X_2 = 1, X_3 = 1)}{\Pr(X_1 = 1, X_2 = 1)} \\
&= \frac{29/48}{17/24} = \frac{29}{34},
\end{aligned}
$$

and that

$$
\begin{aligned}
\Pr(X_3 = 1 \mid X_2 = 1) &= \frac{\Pr(X_2 = 1, X_3 = 1)}{\Pr(X_2 = 1)} \\
&= \frac{17/24}{5/6} = \frac{17}{20}.
\end{aligned}
$$

Hence $\Pr(X_3 = 1 \mid X_2 = 1) \neq \Pr(X_3 = 1 \mid X_1 = 1, X_2 = 1)$; this HMM does not satisfy the Markov property. That some HMMs do satisfy the property, however, is clear. For instance, a two-state Bernoulli–HMM can degenerate in obvious fashion to the underlying Markov chain; one simply identifies each of the two observable values with one of the two underlying states. For the conditions under which an HMM will itself satisfy the Markov property, see Spreij (2001).

2.3.3 The likelihood when data are missing at random

In a time series context it is potentially awkward if some of the data are missing. In the case of hidden Markov time series models, however, the adjustment that needs to be made to the likelihood computation if data are missing turns out to be a simple one.

Suppose, for example, that one has available the observations x_1, x_2, x_4, x_7, x_8, ..., x_T of an HMM, but the observations x_3, x_5 and x_6 are missing at random. Then the likelihood is given by

$$\Pr(X_1 = x_1, X_2 = x_2, X_4 = x_4, X_7 = x_7, \ldots, X_T = x_T)$$
$$= \sum \delta_{c_1} \gamma_{c_1,c_2} \gamma_{c_2,c_4}(2) \gamma_{c_4,c_7}(3) \gamma_{c_7,c_8} \cdots \gamma_{c_{T-1},c_T}$$
$$\times p_{c_1}(x_1) p_{c_2}(x_2) p_{c_4}(x_4) p_{c_7}(x_7) \cdots p_{c_T}(x_T),$$

where (as before) $\gamma_{ij}(k)$ denotes a k-step transition probability, and the sum is taken over all c_t other than c_3, c_5 and c_6. But this is just

$$\sum \delta_{c_1} p_{c_1}(x_1) \gamma_{c_1,c_2} p_{c_2}(x_2) \gamma_{c_2,c_4}(2) p_{c_4}(x_4) \gamma_{c_4,c_7}(3) p_{c_7}(x_7)$$
$$\cdots \times \gamma_{c_{T-1},c_T} p_{c_T}(x_T)$$
$$= \boldsymbol{\delta} \mathbf{P}(x_1) \boldsymbol{\Gamma} \mathbf{P}(x_2) \boldsymbol{\Gamma}^2 \mathbf{P}(x_4) \boldsymbol{\Gamma}^3 \mathbf{P}(x_7) \cdots \boldsymbol{\Gamma} \mathbf{P}(x_T) \mathbf{1}'.$$

With $L_T^{-(3,5,6)}$ denoting the likelihood of the observations other than x_3, x_5 and x_6, our conclusion is therefore that

$$L_T^{-(3,5,6)} = \boldsymbol{\delta} \mathbf{P}(x_1) \boldsymbol{\Gamma} \mathbf{P}(x_2) \boldsymbol{\Gamma}^2 \mathbf{P}(x_4) \boldsymbol{\Gamma}^3 \mathbf{P}(x_7) \cdots \boldsymbol{\Gamma} \mathbf{P}(x_T) \mathbf{1}'.$$

The easiest way to summarize this conclusion is to say that, in the expression for the likelihood, the diagonal matrices $\mathbf{P}(x_t)$ corresponding to missing observations x_t are replaced by the identity matrix; equivalently, the corresponding state-dependent probabilities $p_i(x_t)$ are replaced by 1 for all states i.

The fact that, even in the case of missing observations, the likelihood of an HMM can be easily computed is especially useful in the derivation of conditional distributions, as will be shown in Section 5.1.

2.3.4 The likelihood when observations are interval-censored

Suppose that we wish to fit a Poisson–HMM to a series of counts, some of which are interval-censored. For instance, the exact value of x_t may be known only for $4 \leq t \leq T$, with the information $x_1 \leq 5$, $2 \leq x_2 \leq 3$ and $x_3 > 10$ available about the remaining observations. For simplicity, let us first assume that the Markov chain has only two states. In that case, one replaces the diagonal matrix $\mathbf{P}(x_1)$ in the likelihood expression (2.12) by the matrix

$$\operatorname{diag}(\Pr(X_1 \leq 5 \mid C_1 = 1), \Pr(X_1 \leq 5 \mid C_1 = 2)),$$

and similarly for $\mathbf{P}(x_2)$ and $\mathbf{P}(x_3)$.

More generally, suppose that $a \leq x_t \leq b$, where a may be $-\infty$ (although that is not relevant to the Poisson case), b may be ∞, and the Markov chain has m states. One replaces $\mathbf{P}(x_t)$ in the likelihood by the $m \times m$ diagonal matrix of which the ith diagonal element is $\Pr(a \leq X_t \leq b \mid C_t = i)$. See Exercise 12. It is worth noting that missing data can be regarded as an extreme case of such interval-censoring.

Exercises

1. Consider a stationary two-state Poisson–HMM with parameters

$$\Gamma = \begin{pmatrix} 0.1 & 0.9 \\ 0.4 & 0.6 \end{pmatrix} \quad \text{and} \quad \lambda = (1,3).$$

In each of the following ways, compute the probability that the first three observations from this model are 0, 2, 1.

(a) Consider all possible sequences of states of the Markov chain that could have occurred. Compute the probability of each sequence, and the probability of the observations given each sequence.

(b) Apply the formula

$$\Pr(X_1 = 0, X_2 = 2, X_3 = 1) = \delta \mathbf{P}(0)\Gamma\mathbf{P}(2)\Gamma\mathbf{P}(1)\mathbf{1}',$$

where

$$\mathbf{P}(s) = \begin{pmatrix} \lambda_1^s e^{-\lambda_1}/s! & 0 \\ 0 & \lambda_2^s e^{-\lambda_2}/s! \end{pmatrix} = \begin{pmatrix} 1^s e^{-1}/s! & 0 \\ 0 & 3^s e^{-3}/s! \end{pmatrix}.$$

2. Consider again the model defined in Exercise 1. In that question you were asked to compute $\Pr(X_1 = 0, X_2 = 2, X_3 = 1)$. Now compute $\Pr(X_1 = 0, X_3 = 1)$ in each of the following ways.

(a) Consider all possible sequences of states of the Markov chain that could have occurred. Compute the probability of each sequence, and the probability of the observations given each sequence.

(b) Apply the formula

$$\Pr(X_1 = 0, X_3 = 1) = \delta \mathbf{P}(0)\Gamma\mathbf{I}_2\Gamma\mathbf{P}(1)\mathbf{1}' = \delta \mathbf{P}(0)\Gamma^2\mathbf{P}(1)\mathbf{1}',$$

and check that this probability is equal to your answer in (a).

3. Consider an m-state HMM $\{X_t : t = 1, 2, \ldots\}$, based on a stationary Markov chain with transition probability matrix Γ and stationary distribution $\delta = (\delta_1, \delta_2, \ldots, \delta_m)$, and having (univariate) state-dependent distributions $p_i(x)$. Let μ_i and σ_i^2 denote the mean and variance of the distribution p_i, μ the vector $(\mu_1, \mu_2, \ldots, \mu_m)$, and $\mathbf{M}$ the matrix $\text{diag}(\mu)$.

Derive the following results for the moments of $\{X_t\}$. (Sometimes, but not always, it is useful to express such results in matrix form.)

(a) $E(X_t) = \sum_{i=1}^{m} \delta_i \mu_i = \delta \mu'$.
(b) $E(X_t^2) = \sum_{i=1}^{m} \delta_i(\sigma_i^2 + \mu_i^2)$.
(c) $\text{Var}(X_t) = \sum_{i=1}^{m} \delta_i(\sigma_i^2 + \mu_i^2) - (\delta \mu')^2$.
(d) If $m = 2$, $\text{Var}(X_t) = \delta_1\sigma_1^2 + \delta_2\sigma_2^2 + \delta_1\delta_2(\mu_1 - \mu_2)^2$.

(e) For $k \in \mathbb{N}$, i.e. for positive integers k,
$$\mathrm{E}(X_t X_{t+k}) = \sum_{i=1}^{m} \sum_{j=1}^{m} \delta_i \mu_i \gamma_{ij}(k) \mu_j = \boldsymbol{\delta} \mathbf{M} \boldsymbol{\Gamma}^k \boldsymbol{\mu}'.$$

(f) For $k \in \mathbb{N}$,
$$\rho(k) = \mathrm{Corr}(X_t, X_{t+k}) = \frac{\boldsymbol{\delta} \mathbf{M} \boldsymbol{\Gamma}^k \boldsymbol{\mu}' - (\boldsymbol{\delta} \boldsymbol{\mu}')^2}{\mathrm{Var}(X_t)}.$$

Note that, if the eigenvalues of $\boldsymbol{\Gamma}$ are distinct, this is a linear combination of the kth powers of those eigenvalues.

Timmermann (2000) and Frühwirth-Schnatter (2006, pp. 308–312) are useful references for moments.

4. (Marginal moments and autocorrelation function of a Poisson–HMM: special case of Exercise 3.) Consider a stationary m-state Poisson–HMM $\{X_t : t = 1, 2, \ldots\}$ with transition probability matrix $\boldsymbol{\Gamma}$ and state-dependent means $\boldsymbol{\lambda} = (\lambda_1, \lambda_2, \ldots, \lambda_m)$. Let $\boldsymbol{\delta} = (\delta_1, \delta_2, \ldots, \delta_m)$ be the stationary distribution of the Markov chain. Let $\boldsymbol{\Lambda} = \mathrm{diag}(\boldsymbol{\lambda})$.

Derive the following results.

(a) $\mathrm{E}(X_t) = \boldsymbol{\delta} \boldsymbol{\lambda}'$.
(b) $\mathrm{E}(X_t^2) = \sum_{i=1}^{m} (\lambda_i^2 + \lambda_i)\delta_i = \boldsymbol{\delta} \boldsymbol{\Lambda} \boldsymbol{\lambda}' + \boldsymbol{\delta} \boldsymbol{\lambda}'$.
(c) $\mathrm{Var}(X_t) = \boldsymbol{\delta} \boldsymbol{\Lambda} \boldsymbol{\lambda}' + \boldsymbol{\delta} \boldsymbol{\lambda}' - (\boldsymbol{\delta} \boldsymbol{\lambda}')^2 = \mathrm{E}(X_t) + \boldsymbol{\delta} \boldsymbol{\Lambda} \boldsymbol{\lambda}' - (\boldsymbol{\delta} \boldsymbol{\lambda}')^2 \geq \mathrm{E}(X_t)$.
(d) For $k \in \mathbb{N}$, $\mathrm{E}(X_t X_{t+k}) = \boldsymbol{\delta} \boldsymbol{\Lambda} \boldsymbol{\Gamma}^k \boldsymbol{\lambda}'$.
(e) For $k \in \mathbb{N}$,
$$\rho(k) = \mathrm{Corr}(X_t, X_{t+k}) = \frac{\boldsymbol{\delta} \boldsymbol{\Lambda} \boldsymbol{\Gamma}^k \boldsymbol{\lambda}' - (\boldsymbol{\delta} \boldsymbol{\lambda}')^2}{\boldsymbol{\delta} \boldsymbol{\Lambda} \boldsymbol{\lambda}' + \boldsymbol{\delta} \boldsymbol{\lambda}' - (\boldsymbol{\delta} \boldsymbol{\lambda}')^2}.$$

(f) For the case $m = 2$, $\rho(k) = Aw^k$, where
$$A = \frac{\delta_1 \delta_2 (\lambda_2 - \lambda_1)^2}{\delta_1 \delta_2 (\lambda_2 - \lambda_1)^2 + \boldsymbol{\delta} \boldsymbol{\lambda}'}$$
and $w = 1 - \gamma_{12} - \gamma_{21}$.

5. Consider the three-state Poisson–HMM $\{X_t\}$ with state-dependent means λ_i ($i = 1, 2, 3$) and transition probability matrix
$$\boldsymbol{\Gamma} = \begin{pmatrix} 1/3 & 1/3 & 1/3 \\ 2/3 & 0 & 1/3 \\ 1/2 & 1/2 & 0 \end{pmatrix}.$$

Assume that the Markov chain is stationary.

Show that the autocorrelation $\rho(k) = \mathrm{Corr}(X_t, X_{t+k})$ is given by
$$\frac{(-\frac{1}{3})^k \left\{ 3(-5\lambda_1 - 3\lambda_2 + 8\lambda_3)^2 + 180(\lambda_2 - \lambda_1)^2 \right\} + k(-\frac{1}{3})^{k-1} \left\{ 4(-5\lambda_1 - 3\lambda_2 + 8\lambda_3)(\lambda_2 - \lambda_1) \right\}}{32 \left\{ 15(\lambda_1^2 + \lambda_1) + 9(\lambda_2^2 + \lambda_2) + 8(\lambda_3^2 + \lambda_3) \right\} - (15\lambda_1 + 9\lambda_2 + 8\lambda_3)^2}.$$

Notice that, as a function of k, this (rather tedious!) expression is a linear combination of $(-\frac{1}{3})^k$ and $k(-\frac{1}{3})^{k-1}$. (This is an example of a non-diagonalizable t.p.m. In practice such cases are not likely to be of interest.)

6. We have the general expression

$$L_T = \delta\mathbf{P}(x_1)\mathbf{\Gamma}\mathbf{P}(x_2)\cdots\mathbf{\Gamma}\mathbf{P}(x_T)\mathbf{1}'$$

for the likelihood of an HMM, e.g. of Poisson type. Consider the special case in which the Markov chain degenerates to a sequence of independent random variables, i.e. an independent mixture model.

Show that, in this case, the likelihood simplifies to the expression given in Equation (1.1) for the likelihood of an *independent* mixture.

7. Consider a multiple sum S of the following general form:

$$S = \sum_{i_1=1}^{m}\sum_{i_2=1}^{m}\cdots\sum_{i_T=1}^{m} f_1(i_1)\prod_{t=2}^{T} f_t(i_{t-1}, i_t).$$

For $i_1 = 1, 2, \ldots, m$, define

$$\alpha_1(i_1) \equiv f_1(i_1);$$

and for $r = 1, 2, \ldots, T-1$ and $i_{r+1} = 1, 2, \ldots, m$, define

$$\alpha_{r+1}(i_{r+1}) \equiv \sum_{i_r=1}^{m} \alpha_r(i_r)\, f_{r+1}(i_r, i_{r+1}).$$

That is, the row vector $\boldsymbol{\alpha}_{r+1}$ is defined by, and can be computed as, $\boldsymbol{\alpha}_{r+1} = \boldsymbol{\alpha}_r \mathbf{F}_{r+1}$, where the $m \times m$ matrix $\mathbf{F}_t$ has (i, j) element equal to $f_t(i, j)$.

(a) Show by induction that $\alpha_T(i_T)$ is precisely the sum over all but i_T, i.e. that

$$\alpha_T(i_T) = \sum_{i_1}\sum_{i_2}\cdots\sum_{i_{T-1}} f_1(i_1)\prod_{t=2}^{T} f_t(i_{t-1}, i_t).$$

(b) Hence show that $S = \sum_{i_T}\alpha_T(i_T) = \boldsymbol{\alpha}_T\mathbf{1}' = \boldsymbol{\alpha}_1\mathbf{F}_2\mathbf{F}_3\cdots\mathbf{F}_T\mathbf{1}'$.

(c) Does this result generalize to nonconstant m?

8.(a) In Section 1.3.3 we defined reversibility for a random process, and showed that the stationary Markov chain with the t.p.m. $\mathbf{\Gamma}$ given below is not reversible.

$$\mathbf{\Gamma} = \begin{pmatrix} 1/3 & 1/3 & 1/3 \\ 2/3 & 0 & 1/3 \\ 1/2 & 1/2 & 0 \end{pmatrix}.$$

Now let $\{X_t\}$ be the stationary HMM with $\boldsymbol{\Gamma}$ as above, and having Poisson state-dependent distributions with means 1, 5 and 10; e.g. in state 1 the observation X_t is distributed Poisson with mean 1. By finding the probabilities $\Pr(X_t = 0, X_{t+1} = 1)$ and $\Pr(X_t = 1, X_{t+1} = 0)$, or otherwise, show that $\{X_t\}$ is irreversible.

(b) Show that, if the Markov chain underlying a stationary HMM is reversible, the HMM is also reversible.

(c) Suppose that the Markov chain underlying a stationary HMM is irreversible. Does it follow that the HMM is irreversible?

9. Write a function `pois-HMM.moments(m,lambda,gamma,lag.max=10)` that computes the expectation, variance and autocorrelation function (for lags 0 to `lag.max`) of an m-state stationary Poisson–HMM with t.p.m. `gamma` and state-dependent means `lambda`.

10. Write the three functions listed below, relating to the marginal distribution of an m-state Poisson–HMM with parameters `lambda`, `gamma`, and possibly `delta`. In each case, if `delta` is not specified, the stationary distribution should be used. You can use your function `statdist` (see Exercise 8(b) of Chapter 1) to provide the stationary distribution.

```
dpois.HMM(x, m, lambda, gamma, delta=NULL)
ppois.HMM(x, m, lambda, gamma, delta=NULL)
dpois.HMM(p, m, lambda, gamma, delta=NULL)
```

The function `dpois.HMM` computes the probability function at the arguments specified by the vector `x`, `ppois.HMM` the distribution function, and `ppois.HMM` the inverse distribution function.

11. Consider the function `pois.HMM.generate_sample` in A.2.1 that generates observations from a stationary m-state Poisson–HMM. Test the function by generating a long sequence of observations (10 000, say), and then check whether the sample mean, variance, ACF and relative frequencies correspond to what you expect.

12. Interval-censored observations

(a) Suppose that, in a series of unbounded counts $x_1, \ldots, x_T$, only the observation x_t is interval-censored, and $a \le x_t \le b$, where b may be ∞.

Prove the statement made in Section 2.3.4 that the likelihood of a Poisson–HMM with m states is obtained by replacing $\mathbf{P}(x_t)$ in the expression (2.12) by the $m \times m$ diagonal matrix of which the ith diagonal element is $\Pr(a \le X_t \le b \mid C_t = i)$.

(b) Extend part (a) to allow for any number of interval-censored observations.

CHAPTER 3

Estimation by direct maximization of the likelihood

3.1 Introduction

We saw in Equation (2.12) that the likelihood of an HMM is given by

$$L_T = \Pr\left(\mathbf{X}^{(T)} = \mathbf{x}^{(T)}\right) = \boldsymbol{\delta}\mathbf{P}(x_1)\boldsymbol{\Gamma}\mathbf{P}(x_2)\cdots\boldsymbol{\Gamma}\mathbf{P}(x_T)\mathbf{1}',$$

where $\boldsymbol{\delta}$ is the initial distribution (that of C_1) and $\mathbf{P}(x)$ the $m \times m$ diagonal matrix with ith diagonal element the state-dependent probability or density $p_i(x)$. In principle we can therefore compute $L_T = \boldsymbol{\alpha}_T\mathbf{1}'$ recursively via

$$\boldsymbol{\alpha}_1 = \boldsymbol{\delta}\mathbf{P}(x_1)$$

and

$$\boldsymbol{\alpha}_t = \boldsymbol{\alpha}_{t-1}\boldsymbol{\Gamma}\mathbf{P}(x_t) \quad \text{for } t = 2, 3, \ldots, T.$$

If the Markov chain is assumed to be stationary (in which case $\boldsymbol{\delta} = \boldsymbol{\delta}\boldsymbol{\Gamma}$), we can choose to use instead

$$\boldsymbol{\alpha}_0 = \boldsymbol{\delta}$$

and

$$\boldsymbol{\alpha}_t = \boldsymbol{\alpha}_{t-1}\boldsymbol{\Gamma}\mathbf{P}(x_t) \quad \text{for } t = 1, 2, \ldots, T.$$

We shall first consider the stationary case.

The number of operations involved is of order Tm^2, making the evaluation of the likelihood quite feasible even for large T. Parameter estimation can therefore be performed by numerical maximization of the likelihood with respect to the parameters.

But there are several problems that need to be addressed when the likelihood is computed in this way and maximized numerically in order to estimate parameters. The main problems are numerical underflow, constraints on the parameters, and multiple local maxima in the likelihood function. In this chapter we first discuss how to overcome these problems, in order to arrive at a general strategy for computing MLEs which is easy to implement. Then we discuss the estimation of standard errors for parameters. We defer to the next chapter the EM algorithm, which necessitates some discussion of the forward and backward probabilities.

3.2 Scaling the likelihood computation

In the case of discrete state-dependent distributions, the elements of α_t, being made up of products of probabilities, become progressively smaller as t increases, and are eventually rounded to zero. In fact, with probability 1 the likelihood approaches 0 or ∞ exponentially fast; see Leroux and Puterman (1992). The problem is therefore not confined to the discrete case and underflow; overflow may occur in a continuous case. The remedy is, however, the same for over- and underflow, and we confine our attention to underflow.

Since the likelihood is a product of matrices, not of scalars, it is not possible to circumvent numerical underflow simply by computing the log of the likelihood as the sum of logs of its factors. In this respect the computation of the likelihood of an independent mixture model is simpler than that of an HMM.

To solve the problem, Durbin *et al.* (1998, p. 78) suggest (*inter alia*) a method of computation that relies on the following approximation. Suppose we wish to compute $\log(p+q)$, where $p > q$. Write $\log(p+q)$ as

$$\log p + \log(1 + q/p) = \log p + \log(1 + \exp(\tilde{q} - \tilde{p})),$$

where $\tilde{p} = \log p$ and $\tilde{q} = \log q$. The function $\log(1 + e^x)$ is then approximated by interpolation from a table of its values; apparently quite a small table will give a reasonable degree of accuracy.

We prefer to compute the logarithm of L_T using a strategy of scaling the vector of forward probabilities α_t. Define, for $t = 0, 1, \ldots, T$, the vector

$$\phi_t = \alpha_t / w_t,$$

where $w_t = \sum_i \alpha_t(i) = \alpha_t \mathbf{1}'$.

First we note certain immediate consequences of the definitions of ϕ_t and w_t:

$$
\begin{aligned}
w_0 = \alpha_0 \mathbf{1}' &= \delta \mathbf{1}' = 1; \\
\phi_0 &= \delta; \\
w_t \phi_t &= w_{t-1} \phi_{t-1} \mathbf{B}_t; \\
L_T = \alpha_T \mathbf{1}' &= w_T(\phi_T \mathbf{1}') = w_T.
\end{aligned}
\tag{3.1}
$$

Hence $L_T = w_T = \prod_{t=1}^{T}(w_t/w_{t-1})$. From (3.1) it follows that

$$w_t = w_{t-1}(\phi_{t-1}\mathbf{B}_t\mathbf{1}'),$$

and so we conclude that

$$\log L_T = \sum_{t=1}^{T} \log(w_t/w_{t-1}) = \sum_{t=1}^{T} \log(\phi_{t-1}\mathbf{B}_t\mathbf{1}').$$

The computation of the log-likelihood is summarized below in the form of an algorithm. Note that $\mathbf{\Gamma}$ and $\mathbf{P}(x_t)$ are $m \times m$ matrices, $\mathbf{v}$ and $\boldsymbol{\phi}_t$ are vectors of length m, u is a scalar, and l is the scalar in which the log-likelihood is accumulated.

$$
\begin{aligned}
&\text{set } \boldsymbol{\phi}_0 \leftarrow \boldsymbol{\delta} \text{ and } l \leftarrow 0 \\
&\text{for } t = 1, 2, \ldots, T \\
&\qquad \mathbf{v} \leftarrow \boldsymbol{\phi}_{t-1}\mathbf{\Gamma P}(x_t) \\
&\qquad u \leftarrow \mathbf{v1}' \\
&\qquad l \leftarrow l + \log u \\
&\qquad \boldsymbol{\phi}_t \leftarrow \mathbf{v}/u \\
&\text{return } l
\end{aligned}
$$

The required log-likelihood, $\log L_T$, is then given by the final value of l. This procedure will avoid underflow in many cases. Clearly, variations of the technique are possible: for instance, the scale factor w_t could be chosen instead to be the largest element of the vector being scaled, or the mean of its elements (as opposed to the sum). See A.1.3 (in Appendix A) for an implementation of the above algorithm.

The algorithm is easily modified to compute the log-likelihood without assuming stationarity of the Markov chain. If $\boldsymbol{\delta}$ is the initial distribution, replace the first two lines above by

$$
\text{set } w_1 \leftarrow \boldsymbol{\delta}\mathbf{P}(x_1)\mathbf{1}', \ \boldsymbol{\phi}_1 \leftarrow \boldsymbol{\delta}\mathbf{P}(x_1)/w_1 \text{ and } l \leftarrow \log w_1
$$
$$
\text{for } t = 2, 3, \ldots, T
$$

Of course, if the initial distribution happens to be the stationary distribution, the more general algorithm still applies.

3.3 Maximization of the likelihood subject to constraints

3.3.1 Reparametrization to avoid constraints

The elements of $\mathbf{\Gamma}$ and those of $\boldsymbol{\lambda}$, the vector of state-dependent means in a Poisson–HMM, do not range over the whole of $\mathbb{R}$, the set of all real numbers. Neither therefore should any sensible estimates of the parameters. In particular, the row sums of $\mathbf{\Gamma}$, and any estimate thereof, should equal one. Thus when maximizing the likelihood we have a constrained optimization problem to solve, not an unconstrained one.

Special-purpose software, e.g. NPSOL (Gill et al., 1986) or the corresponding NAG routine E04UCF, can be used to maximize a function of several variables which are subject to constraints. The advice of Gill, Murray and Wright (1981, p. 267) is that it is 'rarely appropriate to alter linearly constrained problems'. However — depending on the implementation and the nature of the data — constrained optimization can be slow. For example, the constrained optimizer constrOptim available in **R** is acknowledged to be slow if the optimum lies on the boundary of the parameter space. We shall focus on the use of the unconstrained optimizer `nlm`. Exercise 3 and A.4 explore the use of `constrOptim`, which can minimize a function subject to linear inequality constraints.

In general, there are two groups of constraints: those that apply to the parameters of the state-dependent distributions and those that apply to the parameters of the Markov chain. The first group of constraints depends on which state-dependent distribution(s) are chosen; e.g. the 'success probability' of a binomial distribution lies between 0 and 1.

In the case of a Poisson–HMM the relevant constraints are:

- the means λ_i of the state-dependent distributions must be nonnegative, for $i = 1, \ldots, m$;

- the rows of the transition probability matrix $\boldsymbol{\Gamma}$ must add to 1, and all the parameters γ_{ij} must be nonnegative.

Here the constraints can be circumvented by making certain transformations. The transformation of the parameters λ_i is relatively easy. Define $\eta_i = \log \lambda_i$, for $i = 1, \ldots, m$. Then $\eta_i \in \mathbb{R}$. After we have maximized the likelihood with respect to the unconstrained parameters, the constrained parameter estimates can be obtained by transforming back: $\hat{\lambda}_i = \exp \hat{\eta}_i$.

The reparametrization of the matrix $\boldsymbol{\Gamma}$ requires more work, but can be accomplished quite elegantly. Note that $\boldsymbol{\Gamma}$ has m^2 entries but only $m(m-1)$ free parameters, as there are m row sum constraints

$$\gamma_{i1} + \gamma_{i2} + \cdots + \gamma_{im} = 1 \quad (i = 1, \ldots, m).$$

We shall show one possible transformation between the m^2 constrained probabilities γ_{ij} and $m(m-1)$ unconstrained real numbers $\tau_{ij}, i \neq j$.

For the sake of readability we show the case $m = 3$. We begin by defining the matrix

$$\mathbf{T} = \begin{pmatrix} - & \tau_{12} & \tau_{13} \\ \tau_{21} & - & \tau_{23} \\ \tau_{31} & \tau_{32} & - \end{pmatrix}, \text{ a matrix with } m(m-1) \text{ entries } \tau_{ij} \in \mathbb{R}.$$

Now let $g : \mathbb{R} \to \mathbb{R}^+$ be a strictly increasing function, e.g.

$$g(x) = e^x \quad \text{or} \quad g(x) = \begin{cases} e^x & x \leq 0 \\ x + 1 & x \geq 0. \end{cases}$$

Define

$$\varrho_{ij} = \begin{cases} g(\tau_{ij}) & \text{for } i \neq j \\ 1 & \text{for } i = j. \end{cases}$$

We then set $\gamma_{ij} = \varrho_{ij} / \sum_{k=1}^{3} \varrho_{ik}$ (for $i, j = 1, 2, 3$) and $\boldsymbol{\Gamma} = (\gamma_{ij})$. It is left to the reader as an exercise to verify that the resulting matrix $\boldsymbol{\Gamma}$ satisfies the constraints of a transition probability matrix. We shall refer to the parameters η_i and τ_{ij} as **working parameters**, and to the parameters λ_i and γ_{ij} as **natural parameters**.

Using the above transformations of $\boldsymbol{\Gamma}$ and $\boldsymbol{\lambda}$, we can perform the calculation of the likelihood-maximizing parameters in two steps.

1. Maximize L_T with respect to the working parameters $\mathbf{T} = \{\tau_{ij}\}$ and $\boldsymbol{\eta} = (\eta_1, \ldots, \eta_m)$. These are all unconstrained.

2. Transform the estimates of the working parameters to estimates of the natural parameters:

$$\hat{\mathbf{T}} \rightarrow \hat{\boldsymbol{\Gamma}}, \; \hat{\boldsymbol{\eta}} \rightarrow \hat{\boldsymbol{\lambda}}.$$

See A.1.1 and A.1.2 for $\mathbf{R}$ functions that transform natural parameters to working and vice versa.

As an illustration we consider the first row of $\boldsymbol{\Gamma}$ for the case $g(x) = e^x$ and $m = 3$. We have

$$
\begin{aligned}
\gamma_{11} &= 1/\left(1 + \exp(\tau_{12}) + \exp(\tau_{13})\right), \\
\gamma_{12} &= \exp(\tau_{12})/\left(1 + \exp(\tau_{12}) + \exp(\tau_{13})\right), \\
\gamma_{13} &= \exp(\tau_{13})/\left(1 + \exp(\tau_{12}) + \exp(\tau_{13})\right).
\end{aligned}
$$

The transformation in the opposite direction is

$$
\begin{aligned}
\tau_{12} &= \log\left(\gamma_{12}/\left(1 - \gamma_{12} - \gamma_{13}\right)\right) = \log\left(\gamma_{12}/\gamma_{11}\right), \\
\tau_{13} &= \log\left(\gamma_{13}/\left(1 - \gamma_{12} - \gamma_{13}\right)\right) = \log\left(\gamma_{13}/\gamma_{11}\right).
\end{aligned}
$$

This generalization of the logit and inverse logit transforms has long been used in (for example) the context of compositional data: see Aitchison (1982), where several other transforms are described as well.

3.3.2 Embedding in a continuous-time Markov chain

A different reparametrization is described by Zucchini and MacDonald (1998). In a continuous-time Markov chain on a finite state-space, the transition probability matrix $\mathbf{P}_t$ is given by $\mathbf{P}_t = \exp(t\mathbf{Q})$, where $\mathbf{Q}$ is the matrix of transition intensities. The row sums of $\mathbf{Q}$ are zero, but the only constraint on the off-diagonal elements of $\mathbf{Q}$ is that they be non-negative. The one-step transition probabilities in a discrete-time Markov chain can therefore be parametrized via $\boldsymbol{\Gamma} = \exp(\mathbf{Q})$. This is effectively the parametrization used in the $\mathbf{R}$ package msm (Jackson et al., 2003).

3.4 Other problems

3.4.1 Multiple maxima in the likelihood

The likelihood of an HMM is a complicated function of the parameters and frequently has several local maxima. The goal of course is to find the global maximum, but there is no simple method of determining in general whether a numerical maximization algorithm has reached the global maximum. Depending on the starting values, it can easily happen that the algorithm identifies a local, but not the global, maximum. This applies also to the main alternative method of estimation, the EM algorithm, which is discussed in Chapter 4. A sensible strategy is therefore to use a range of starting values for the maximization, and to see whether the same maximum is identified in each case.

3.4.2 Starting values for the iterations

It is often easy to find plausible starting values for some of the parameters of an HMM: for instance, if one seeks to fit a Poisson–HMM with two states, and the sample mean is 10, one could try 8 and 12, or 5 and 15, for the values of the two state-dependent means. More systematic strategies based on the quantiles of the observations are possible, however: e.g. if the model has three states, use as the starting values of the state-dependent means the lower quartile, median and upper quartile of the observed counts.

It is less easy to guess values of the transition probabilities γ_{ij}. One strategy is to assign a common starting value (e.g. 0.01 or 0.05) to all the off-diagonal transition probabilities. A consequence of such a choice, perhaps convenient, is that the corresponding stationary distribution is uniform over the states; this follows by symmetry. Choosing good starting values for parameters tends to steer one away from numerical instability.

3.4.3 Unbounded likelihood

In the case of HMMs with continuous state-dependent distributions, just as in the case of independent mixtures (see Section 1.2.3), it may happen that the likelihood is unbounded in the vicinity of certain parameter combinations. As before, we suggest that, if this creates difficulties, one maximizes the discrete likelihood instead of the joint density. This has the advantage in any case that it applies more generally to data known only up to some interval. Applications of this kind are described in Sections 10.3 and 10.4; code for the latter is in A.3.

3.5 Example: earthquakes

Figure 3.1 shows the result of fitting (stationary) Poisson–hidden Markov models with two and three states to the earthquakes series by means of the unconstrained optimizer nlm. The relevant code appears in A.1. The

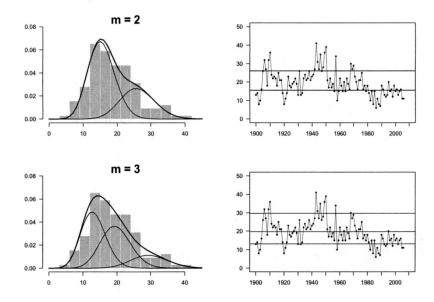

Figure 3.1 *Earthquakes series. Left: marginal distributions of Poisson–HMMs with 2 and 3 states, and their components, compared with a histogram of the observations. Right: the state-dependent means (horizontal lines) compared to the observations.*

two-state model is

$$\boldsymbol{\Gamma} = \left(\begin{array}{cc} 0.9340 & 0.0660 \\ 0.1285 & 0.8715 \end{array} \right),$$

with $\boldsymbol{\delta} = (0.6608, 0.3392)$, $\boldsymbol{\lambda} = (15.472, 26.125)$, and log-likelihood given by $l = -342.3183$. It is clear that the fitted (Markov-dependent) mixture of two Poisson distributions provides a much better fit to the marginal distribution of the observations than does a single Poisson distribution, but the fit can be further improved by using a mixture of three or four Poisson distributions.

The three-state model is

$$\boldsymbol{\Gamma} = \left(\begin{array}{ccc} 0.955 & 0.024 & 0.021 \\ 0.050 & 0.899 & 0.051 \\ 0.000 & 0.197 & 0.803 \end{array} \right),$$

with $\boldsymbol{\delta} = (0.4436, 0.4045, 0.1519)$, $\boldsymbol{\lambda} = (13.146, 19.721, 29.714)$ and $l =$

-329.4603, and the four-state is as follows:

$$\Gamma = \begin{pmatrix} 0.805 & 0.102 & 0.093 & 0.000 \\ 0.000 & 0.976 & 0.000 & 0.024 \\ 0.050 & 0.000 & 0.902 & 0.048 \\ 0.000 & 0.000 & 0.188 & 0.812 \end{pmatrix},$$

with $\delta = (0.0936, 0.3983, 0.3643, 0.1439)$, $\lambda = (11.283, 13.853, 19.695, 29.700)$, and $l = -327.8316$.

The means and variances of the marginal distributions of the four models compare as follows with those of the observations. By a one-state Poisson–HMM we mean a model that assumes that the observations are realizations of independent Poisson random variables with common mean.

	mean	variance
observations:	19.364	51.573
'one-state HMM':	19.364	19.364
two-state HMM:	19.086	44.523
three-state HMM:	18.322	50.709
four-state HMM:	18.021	49.837

As regards the autocorrelation functions of the models, i.e. $\rho(k) = \text{Corr}(X_{t+k}, X_t)$, we have the following results, valid for all $k \in \mathbb{N}$, based on the conclusions of Exercise 4 of Chapter 2:

- two states: $\rho(k) = 0.5713 \times 0.8055^k$;
- three states: $\rho(k) = 0.4447 \times 0.9141^k + 0.1940 \times 0.7433^k$;
- four states: $\rho(k) = 0.2332 \times 0.9519^k + 0.3682 \times 0.8174^k + 0.0369 \times 0.7252^k$.

In all these cases the ACF is just a linear combination of the kth powers of the eigenvalues other than 1 of the transition probability matrix.

For model selection, e.g. choosing between competing models such as HMMs and independent mixtures, or choosing the number of components in either, see Section 6.1.

A phenomenon that is noticeable when one fits models with three or more states to relatively short series is that the estimates of one or more of the transition probabilities turn out to be very close to zero; see the three-state model above (one such probability, γ_{13}) and the four-state model (six of the twelve off-diagonal transition probabilities).

This phenomenon can be explained as follows. In a stationary Markov chain, the expected number of transitions from state i to state j in a series of T observations is $(T-1)\delta_i\gamma_{ij}$. For $\delta_3 = 0.152$ and $T = 107$ (as in our three-state model), this expectation will be less than 1 if $\gamma_{31} < 0.062$. In such a series, therefore, it is likely that if γ_{31} is fairly small there will be no transitions from state 3 to state 1, and so when we seek to estimate γ_{31} in an HMM the estimate is likely to be effectively zero. As m increases, the probabilities δ_i and γ_{ij} get smaller on average; this makes it increasingly likely that at least one estimated transition probability is effectively zero.

3.6 Standard errors and confidence intervals

Relatively little is known about the properties of the maximum likelihood estimators of HMMs; only asymptotic results are available. To exploit these results one requires estimates of the variance-covariance matrix of the estimators of the parameters. One can estimate the standard errors from the Hessian of the log-likelihood at the maximum, but this approach runs into difficulties when some of the parameters are on the boundary of their parameter space, which occurs quite often when HMMs are fitted. An alternative here is the parametric bootstrap, for which see Section 3.6.2. The algorithm is easy to code (see A.2.1), but the computations are time-consuming.

3.6.1 Standard errors via the Hessian

Although the point estimates $\widehat{\Theta} = (\widehat{\Gamma}, \widehat{\lambda})$ are easy to compute, exact interval estimates are not available. Cappé *et al.* (2005, Chapter 12) show that, under certain regularity conditions, the MLEs of HMM parameters are consistent, asymptotically normal and efficient. Thus, if we can estimate the standard errors of the MLEs, then, using the asymptotic normality, we can also compute approximate confidence intervals. However, as pointed out by Frühwirth-Schnatter (2006, p. 53) in the context of independent mixture models, 'The regularity conditions are often violated, including cases of great practical concern, among them small data sets, mixtures with small component weights, and overfitting mixtures with too many components.' Furthermore, McLachlan and Peel (2000, p. 68) warn: 'In particular for mixture models, it is well known that the sample size n has to be very large before the asymptotic theory of maximum likelihood applies.'

With the above caveats in mind we can, to estimate the standard errors of the MLEs of an HMM, use the approximate Hessian of minus the log-likelihood at the minimum, e.g. as supplied by nlm. We can invert it and so estimate the asymptotic variance-covariance matrix of the estimators of the parameters. A problem with this suggestion is that, if the parameters have been transformed, the Hessian available will be that which refers to the working parameters ϕ_i, not the original, more readily interpretable, natural parameters θ_i (Γ and λ in the case of a Poisson–HMM).

The situation is therefore that we have the Hessian

$$\mathbf{H} = -\left(\frac{\partial^2 l}{\partial \phi_i \partial \phi_j}\right)$$

available at the minimum of $-l$, and what we really need is the Hessian with respect to the natural parameters:

$$\mathbf{G} = -\left(\frac{\partial^2 l}{\partial \theta_i \partial \theta_j}\right).$$

There is, however, the following relationship between the two Hessians at the minimum:

$$\mathbf{H} = \mathbf{MGM}' \quad \text{and} \quad \mathbf{G}^{-1} = \mathbf{M}'\mathbf{H}^{-1}\mathbf{M}, \tag{3.2}$$

where $\mathbf{M}$ is defined by $m_{ij} = \partial \theta_j / \partial \phi_i$. (Note that all the derivatives appearing here are as evaluated at the minimum.) In the case of a Poisson–HMM, the elements of $\mathbf{M}$ are quite simple: see Exercise 7 for details.

With $\mathbf{M}$ at our disposal, we can use (3.2) to deduce $\mathbf{G}^{-1}$ from $\mathbf{H}^{-1}$, and use $\mathbf{G}^{-1}$ to find standard errors for the natural parameters, provided such parameters are not on the boundary of the parameter space. It is however true in many applications that some of the estimated parameters lie on or very close to the boundary; this limits the usefulness of the above results.

As already pointed out on p. 52, for series of moderate length the estimates of some transition probabilities are expected to be close to zero. This is true of $\hat{\gamma}_{13}$ in the three-state model for the earthquakes series. An additional example of this type can be found in Section 10.2.2. In Section 12.2.1, several of the estimates of the parameters in the state-dependent distributions are practically zero, their lower bound; see Table 12.1. The same phenomenon is apparent in Section 16.9.2; see Table 16.1.

Recursive computation of the Hessian

An alternative method of computing the Hessian is that of Lystig and Hughes (2002). They present the forward algorithm $\boldsymbol{\alpha}_t = \boldsymbol{\alpha}_{t-1}\boldsymbol{\Gamma}\mathbf{P}(x_t)$ in a form which incorporates automatic or 'natural' scaling, and then extend that approach in order to compute (in a single pass, along with the log-likelihood) its Hessian and gradient with respect to the natural parameters, those we have denoted above by θ_i. Turner (2008) has used this approach in order to find the analytical derivatives needed to maximize HMM likelihoods directly by the Levenberg–Marquardt algorithm.

While this may be a more efficient and more accurate method of computing the Hessian than that outlined above, it does not solve the fundamental problem that the use of the Hessian to compute standard errors (and thence confidence intervals) is unreliable if some of the parameters are on or near the boundary of their parameter space.

3.6.2 Bootstrap standard errors and confidence intervals

As an alternative to the technique described in Section 3.6.1 one may use the **parametric bootstrap** (Efron and Tibshirani, 1993). Roughly speaking, the idea of the parametric bootstrap is to assess the properties of the model with parameters Θ by using those of the model with parameters $\widehat{\Theta}$. The following steps are performed to estimate the variance-covariance matrix of $\widehat{\Theta}$.

1. Fit the model, i.e. compute $\widehat{\Theta}$.

2.(a) Generate a sample, called a bootstrap sample, of observations from the fitted model, i.e. the model with parameters $\widehat{\Theta}$. The length should be the same as the original number of observations.

   (b) Estimate the parameters Θ by $\widehat{\Theta}^*$ for the bootstrap sample.

   (c) Repeat steps (a) and (b) B times (with B 'large') and record the values $\widehat{\Theta}^*$.

The variance-covariance matrix of $\widehat{\Theta}$ is then estimated by the sample variance-covariance matrix of the bootstrap estimates $\widehat{\Theta}^*(b)$, $b = 1, 2, \ldots, B$:

$$\widehat{\text{Var-Cov}}(\widehat{\Theta}) = \frac{1}{B-1} \sum_{b=1}^{B} \left(\widehat{\Theta}^*(b) - \widehat{\Theta}^*(.) \right)' \left(\widehat{\Theta}^*(b) - \widehat{\Theta}^*(.) \right),$$

where $\widehat{\Theta}^*(.) = B^{-1} \sum_{b=1}^{B} \widehat{\Theta}^*(b)$.

The parametric bootstrap requires code to generate realizations from a fitted model; for a Poisson–HMM this is given in A.2.1. Since code to fit models is available, that same code can be used to fit models to the bootstrap sample.

The bootstrap method can be used to estimate confidence intervals directly. In the example given in the next section we use the well-known 'percentile method' (Efron and Tibshirani, 1993); other options are available.

3.7 Example: the parametric bootstrap applied to the three-state model for the earthquakes data

A bootstrap sample of size 500 was generated from the three-state model for the earthquakes data, which appears on p. 51. In fitting models to the bootstrap samples, we noticed that, in two cases out of the 500, the starting values we were in general using caused numerical instability or

Table 3.1 *Earthquakes data: bootstrap confidence intervals for the parameters of the three-state HMM.*

parameter	MLE	90% conf. limits	
λ_1	13.146	11.463	14.253
λ_2	19.721	13.708	21.142
λ_3	29.714	20.929	33.160
γ_{11}	0.954	0.750	0.988
γ_{12}	0.024	0.000	0.195
γ_{13}	0.021	0.000	0.145
γ_{21}	0.050	0.000	0.179
γ_{22}	0.899	0.646	0.974
γ_{23}	0.051	0.000	0.228
γ_{31}	0.000	0.000	0.101
γ_{32}	0.197	0.000	0.513
γ_{33}	0.803	0.481	0.947
δ_1	0.444	0.109	0.716
δ_2	0.405	0.139	0.685
δ_3	0.152	0.042	0.393

Table 3.2 *Earthquakes data: bootstrap estimates of the correlations of the estimators of λ_i, for $i = 1, 2, 3$.*

	λ_1	λ_2	λ_3
λ_1	1.000	0.483	0.270
λ_2		1.000	0.688
λ_3			1.000

convergence problems. By choosing better starting values for these two cases we were able to fit models successfully and complete the exercise. The resulting sample of parameter values then produced the 90% confidence intervals for the parameters that are displayed in Table 3.1, and the estimated parameter correlations that are displayed in Table 3.2. What is noticeable is that the intervals for the state-dependent means λ_i overlap, the intervals for the stationary probabilities δ_i are very wide, and the estimators $\hat{\lambda}_i$ are quite strongly correlated.

These results, in particular the correlations shown in Table 3.2, should make one wary of over-interpreting a model with nine parameters based on only 107 (dependent) observations. In particular, they suggest that the states are not well defined, and one should be cautious of attaching a substantive interpretation to them.

Exercises

1. Consider the following parametrization of the t.p.m. of an m-state
   Markov chain. Let $\tau_{ij} \in \mathbb{R}$ $(i, j = 1, 2, \ldots, m; i \neq j)$ be $m(m-1)$
   arbitrary real numbers. Let $g : \mathbb{R} \to \mathbb{R}^+$ be some strictly increasing
   function, e.g. $g(x) = e^x$. Define ϱ_{ij} and γ_{ij} as on p. 48.

   (a) Show that the matrix $\boldsymbol{\Gamma}$ with entries γ_{ij} that are constructed in
   this way is a t.p.m., i.e. show that $0 \leq \gamma_{ij} \leq 1$ for all i and j, and
   that the row sums of $\boldsymbol{\Gamma}$ are equal to one.

   (b) Given a t.p.m. $\boldsymbol{\Gamma} = \{\gamma_{ij} : i, j = 1, 2, \ldots, m\}$, derive an expression
   for the parameters τ_{ij}, for $i, j = 1, 2, \ldots, m;\ i \neq j$.

2. The purpose of this exercise is to investigate the numerical behaviour
   of an 'unscaled' evaluation of the likelihood of an HMM, and to compare this with the behaviour of an alternative algorithm that applies
   scaling.

   Consider the stationary two-state Poisson–HMM with parameters

   $$\boldsymbol{\Gamma} = \begin{pmatrix} 0.9 & 0.1 \\ 0.2 & 0.8 \end{pmatrix}, \qquad (\lambda_1, \lambda_2) = (1, 5).$$

   Compute the likelihood, L_{10}, of the following sequence of ten observations in two ways: 2, 8, 6, 3, 6, 1, 0, 0, 4, 7.

   (a) Use the unscaled method $L_{10} = \boldsymbol{\alpha}_{10} \mathbf{1}'$, where $\boldsymbol{\alpha}_0 = \boldsymbol{\delta}$ and $\boldsymbol{\alpha}_t = \boldsymbol{\alpha}_{t-1} \mathbf{B}_t$;

   $$\mathbf{B}_t = \boldsymbol{\Gamma} \begin{pmatrix} p_1(x_t) & 0 \\ 0 & p_2(x_t) \end{pmatrix};$$

   and

   $$p_i(x_t) = \lambda_i^{x_t} e^{-\lambda_i} / x_t!, \quad i = 1, 2; \quad t = 1, 2, \ldots, 10.$$

   Examine the numerical values of the vectors $\boldsymbol{\alpha}_0, \boldsymbol{\alpha}_1, \ldots, \boldsymbol{\alpha}_{10}$.

   (b) Use the algorithm given in Section 3.2 to compute $\log L_{10}$.
   Examine the numerical values of the vectors $\boldsymbol{\phi}_0, \boldsymbol{\phi}_1, \ldots, \boldsymbol{\phi}_{10}$. (It is
   easiest to store these vectors as rows in an 11×2 matrix.)

3. Use the **R** function constrOptim to fit HMMs with two to four states
   to the earthquakes data, and compare your models with those given
   in Section 3.5.

4. Consider the following transformation:

   $$\begin{aligned}
   w_1 &= \sin^2 \theta_1 \\
   w_i &= \left(\Pi_{j=1}^{i-1} \cos^2 \theta_j \right) \sin^2 \theta_i \qquad i = 2, \ldots, m-1 \\
   w_m &= \Pi_{i=1}^{m-1} \cos^2 \theta_i.
   \end{aligned}$$

Show how this transformation can be used to convert the constraints

$$\sum_{i=1}^{m} w_i = 1, \qquad w_i \geq 0 \qquad i = 1, \ldots, m$$

into simple 'box constraints', i.e. constraints of the form $a \leq \theta_i \leq b$. How could this be used in the context of estimation in HMMs?

5.(a) Consider a stationary Markov chain, with t.p.m. $\boldsymbol{\Gamma}$ and stationary distribution $\boldsymbol{\delta}$. Show that the expected number of transitions from state i to state j in a series of T observations (i.e. in $T-1$ transitions) is $(T-1)\delta_i\gamma_{ij}$.

Hint: this expectation is $\sum_{t=2}^{T} \Pr(X_{t-1} = i, X_t = j)$.

(b) Show that, for $\delta_3 = 0.152$ and $T = 107$, this expectation is less than 1 if $\gamma_{31} < 0.062$.

6. Prove the relation (3.2) between the Hessian $\mathbf{H}$ of $-l$ with respect to the working parameters and the Hessian $\mathbf{G}$ of $-l$ with respect to the natural parameters, both being evaluated at the minimum of $-l$.

7. (See Section 3.6.1.) Consider an m-state Poisson–HMM, with natural parameters γ_{ij} and λ_i, and working parameters τ_{ij} and η_i defined as in Section 3.3.1, with $g(x) = e^x$.

(a) Show that

$$\partial\gamma_{ij}/\partial\tau_{ij} = \gamma_{ij}(1 - \gamma_{ij}), \text{ for all } i, j; \quad \partial\gamma_{ij}/\partial\tau_{il} = -\gamma_{ij}\gamma_{il}, \text{ for } j \neq l;$$
$$\partial\gamma_{ij}/\partial\tau_{kl} = 0, \text{ for } i \neq k; \text{ and} \qquad \partial\lambda_i/\partial\eta_i = e^{\eta_i} = \lambda_i, \text{ for all } i.$$

(b) Hence find the matrix $\mathbf{M}$ in this case.

8. Modify the **R** code in A.1 in order to fit a Poisson–HMM to interval-censored observations. (Assume that the observations are available as a $T \times 2$ matrix of which the first column contains the lower bound of the observation and the second the upper bound, possibly Inf.)

9. Verify the autocorrelation functions given on p. 52 for the two-, three- and four-state models for the earthquakes data. (Hint: use the **R** function eigen to find the eigenvalues and -vectors of the relevant transition probability matrices.)

10. Consider again the soap sales series introduced in Exercise 5 of Chapter 1.

(a) Fit stationary Poisson–HMMs with two, three and four states to these data.

(b) Find the marginal means and variances, and the ACFs, of these models, and compare them with their sample equivalents.

CHAPTER 4

Estimation by the EM algorithm

I know many statisticians are deeply in love with the EM algorithm [...]

Speed (2008)

A commonly used method of fitting HMMs is the EM algorithm, which we shall describe in Section 4.2, the crux of this chapter. The tools we need to do so are the forward and the backward probabilities, which are also used for decoding and state prediction in Chapter 5. In establishing some useful propositions concerning the forward and backward probabilities we invoke several properties of HMMs which are fairly obvious given the structure of an HMM; we defer the proofs of such properties to Appendix B.

In the context of HMMs the EM algorithm is known as the Baum–Welch algorithm. The Baum–Welch algorithm is designed to estimate the parameters of an HMM whose Markov chain is homogeneous but not necessarily stationary. Thus, in addition to the parameters of the state-dependent distributions and the t.p.m. $\boldsymbol{\Gamma}$, the initial distribution $\boldsymbol{\delta}$ is also estimated; it is not assumed that $\boldsymbol{\delta\Gamma} = \boldsymbol{\delta}$. Indeed the method has to be modified if this assumption is made; see Section 4.2.5.

4.1 Forward and backward probabilities

In Section 2.3.2 we have, for $t = 1, 2, \ldots, T$, defined the (row) vector $\boldsymbol{\alpha}_t$ as follows:

$$\boldsymbol{\alpha}_t = \boldsymbol{\delta}\mathbf{P}(x_1)\boldsymbol{\Gamma}\mathbf{P}(x_2)\cdots\boldsymbol{\Gamma}\mathbf{P}(x_t) = \boldsymbol{\delta}\mathbf{P}(x_1)\prod_{s=2}^{t}\boldsymbol{\Gamma}\mathbf{P}(x_s), \qquad (4.1)$$

with $\boldsymbol{\delta}$ denoting the initial distribution of the Markov chain. We have referred to the elements of $\boldsymbol{\alpha}_t$ as **forward probabilities**, but we have given no reason even for their description as probabilities. One of the purposes of this section is to show that $\alpha_t(j)$, the jth component of $\boldsymbol{\alpha}_t$, is indeed a probability, the joint probability $\Pr(X_1 = x_1, X_2 = x_2, \ldots, X_t = x_t, C_t = j)$.

We shall also need the vector of **backward probabilities** $\boldsymbol{\beta}_t$ which, for $t = 1, 2, \ldots, T$, is defined by

$$\beta_t' = \mathbf{\Gamma P}(x_{t+1}) \mathbf{\Gamma P}(x_{t+2}) \cdots \mathbf{\Gamma P}(x_T) \mathbf{1}' = \left(\prod_{s=t+1}^{T} \mathbf{\Gamma P}(x_s) \right) \mathbf{1}', \quad (4.2)$$

with the convention that an empty product is the identity matrix; the case $t = T$ therefore yields $\beta_T = \mathbf{1}$. We shall show that $\beta_t(j)$, the j th component of β_t, can be identified as the *conditional* probability $\Pr(X_{t+1} = x_{t+1}, X_{t+2} = x_{t+2}, \ldots, X_T = x_T \mid C_t = j)$. It will then follow that, for $t = 1, 2, \ldots, T$,

$$\alpha_t(j)\beta_t(j) = \Pr(\mathbf{X}^{(T)} = \mathbf{x}^{(T)}, C_t = j).$$

4.1.1 Forward probabilities

It follows immediately from the definition of α_t that, for $t = 1, 2, \ldots, T-1$, $\alpha_{t+1} = \alpha_t \mathbf{\Gamma P}(x_{t+1})$ or, in scalar form,

$$\alpha_{t+1}(j) = \left(\sum_{i=1}^{m} \alpha_t(i)\gamma_{ij} \right) p_j(x_{t+1}). \quad (4.3)$$

We shall now use the above recursion, and Equation (B.1) in Appendix B, to prove the following result by induction.

Proposition 2 *For $t = 1, 2, \ldots, T$ and $j = 1, 2, \ldots, m$,*

$$\alpha_t(j) = \Pr(\mathbf{X}^{(t)} = \mathbf{x}^{(t)}, C_t = j).$$

Proof. Since $\alpha_1 = \delta \mathbf{P}(x_1)$, we have

$$\alpha_1(j) = \delta_j\, p_j(x_1) = \Pr(C_1 = j) \Pr(X_1 = x_1 \mid C_1 = j),$$

hence $\alpha_1(j) = \Pr(X_1 = x_1, C_1 = j)$; i.e. the proposition holds for $t = 1$. We now show that, if the proposition holds for some $t \in \mathbb{N}$, then it also holds for $t + 1$.

$$
\begin{aligned}
\alpha_{t+1}(j) &= \sum_{i=1}^{m} \alpha_t(i)\gamma_{ij}p_j(x_{t+1}) \quad \text{(see (4.3))} \\
&= \sum_i \Pr(\mathbf{X}^{(t)} = \mathbf{x}^{(t)}, C_t = i) \Pr(C_{t+1} = j \mid C_t = i) \\
&\qquad\qquad \times \Pr(X_{t+1} = x_{t+1} \mid C_{t+1} = j) \\
&= \sum_i \Pr(\mathbf{X}^{(t+1)} = \mathbf{x}^{(t+1)}, C_t = i, C_{t+1} = j) \quad (4.4) \\
&= \Pr(\mathbf{X}^{(t+1)} = \mathbf{x}^{(t+1)}, C_{t+1} = j),
\end{aligned}
$$

as required. The crux is the line numbered (4.4); Equation (B.1) provides the justification thereof. $\qquad\square$

4.1.2 Backward probabilities

It follows immediately from the definition of β_t that $\beta'_t = \Gamma \mathbf{P}(x_{t+1})\beta'_{t+1}$, for $t = 1, 2, \ldots, T - 1$.

Proposition 3 *For $t = 1, 2, \ldots, T - 1$ and $i = 1, 2, \ldots, m$,*

$$\beta_t(i) = \Pr(X_{t+1} = x_{t+1}, X_{t+2} = x_{t+2}, \ldots, X_T = x_T \mid C_t = i),$$

provided that $\Pr(C_t = i) > 0$. In a more compact notation:

$$\beta_t(i) = \Pr(\mathbf{X}^T_{t+1} = \mathbf{x}^T_{t+1} \mid C_t = i),$$

where $\mathbf{X}^b_a$ denotes the vector $(X_a, X_{a+1}, \ldots, X_b)$.

This proposition identifies $\beta_t(i)$ as a conditional probability: the probability of the observations being $x_{t+1}, \ldots, x_T$, given that the Markov chain is in state i at time t. (Recall that the forward probabilities are joint probabilities, not conditional.)

Proof. The proof is by induction, but essentially comes down to Equations (B.5) and (B.6) of Appendix B. These are

$$\Pr(X_{t+1} \mid C_{t+1}) \Pr(\mathbf{X}^T_{t+2} \mid C_{t+1}) = \Pr(\mathbf{X}^T_{t+1} \mid C_{t+1}), \tag{B.5}$$

and

$$\Pr(\mathbf{X}^T_{t+1} \mid C_{t+1}) = \Pr(\mathbf{X}^T_{t+1} \mid C_t, C_{t+1}). \tag{B.6}$$

To establish validity for $T = t - 1$, note that, since $\beta'_{T-1} = \Gamma \mathbf{P}(x_T)\mathbf{1}'$,

$$\beta_{T-1}(i) = \sum_j \Pr(C_T = j \mid C_{T-1} = i) \Pr(X_T = x_T \mid C_T = j). \tag{4.5}$$

But, by (B.6),

$$\begin{aligned}
\Pr(C_T \mid C_{T-1}) \Pr(X_T \mid C_T) &= \Pr(C_T \mid C_{T-1}) \Pr(X_T \mid C_{T-1}, C_T) \\
&= \Pr(X_T, C_{T-1}, C_T) / \Pr(C_{T-1}). \tag{4.6}
\end{aligned}$$

Substitute from (4.6) into (4.5), and the result is

$$\begin{aligned}
\beta_{T-1}(i) &= \frac{1}{\Pr(C_{T-1} = i)} \sum_j \Pr(X_T = x_T, C_{T-1} = i, C_T = j) \\
&= \Pr(X_T = x_T, C_{T-1} = i) / \Pr(C_{T-1} = i) \\
&= \Pr(X_T = x_T \mid C_{T-1} = i),
\end{aligned}$$

as required.

To show that validity for $t + 1$ implies validity for t, first note that the recursion for β_t, and the inductive hypothesis, establish that

$$\beta_t(i) = \sum_j \gamma_{ij} \Pr(X_{t+1} = x_{t+1} \mid C_{t+1} = j) \Pr(\mathbf{X}^T_{t+2} = \mathbf{x}^T_{t+2} \mid C_{t+1} = j). \tag{4.7}$$

But (B.5) and (B.6) imply that

$$\Pr(X_{t+1} \mid C_{t+1}) \Pr(\mathbf{X}_{t+2}^T \mid C_{t+1}) = \Pr(\mathbf{X}_{t+1}^T \mid C_t, C_{t+1}). \qquad (4.8)$$

Substitute from (4.8) into (4.7), and the result is

$$
\begin{aligned}
\beta_t(i) &= \sum_j \Pr(C_{t+1} = j \mid C_t = i) \Pr(\mathbf{X}_{t+1}^T = \mathbf{x}_{t+1}^T \mid C_t = i, C_{t+1} = j) \\
&= \frac{1}{\Pr(C_t = i)} \sum_j \Pr(\mathbf{X}_{t+1}^T = \mathbf{x}_{t+1}^T, C_t = i, C_{t+1} = j) \\
&= \Pr(\mathbf{X}_{t+1}^T = \mathbf{x}_{t+1}^T, C_t = i) / \Pr(C_t = i),
\end{aligned}
$$

which is the required conditional probability. $\qquad\square$

Note that the backward probabilities require a backward pass through the data for their evaluation, just as the forward probabilities require a forward pass; hence the names.

4.1.3 Properties of forward and backward probabilities

We now establish a result relating the forward and backward probabilities $\alpha_t(i)$ and $\beta_t(i)$ to the probabilities $\Pr(\mathbf{X}^{(T)} = \mathbf{x}^{(T)}, C_t = i)$. This we shall use in applying the EM algorithm to HMMs, and in local decoding: see Section 5.3.1.

Proposition 4 *For $t = 1, 2, \ldots, T$ and $i = 1, 2, \ldots, m$,*

$$\alpha_t(i)\beta_t(i) = \Pr(\mathbf{X}^{(T)} = \mathbf{x}^{(T)}, C_t = i), \qquad (4.9)$$

and consequently $\boldsymbol{\alpha}_t \boldsymbol{\beta}_t' = \Pr(\mathbf{X}^{(T)} = \mathbf{x}^{(T)}) = L_T$, for each such t.

Proof. By the preceding two propositions,

$$
\begin{aligned}
\alpha_t(i)\beta_t(i) &= \Pr(\mathbf{X}_1^t, C_t = i) \Pr(\mathbf{X}_{t+1}^T \mid C_t = i) \\
&= \Pr(C_t = i) \Pr(\mathbf{X}_1^t \mid C_t = i) \Pr(\mathbf{X}_{t+1}^T \mid C_t = i).
\end{aligned}
$$

Now apply the conditional independence of $\mathbf{X}_1^t$ and $\mathbf{X}_{t+1}^T$ given C_t (see Equation (B.7) of Appendix B), and the result is that

$$\alpha_t(i)\beta_t(i) = \Pr(C_t = i) \Pr(\mathbf{X}_1^t, \mathbf{X}_{t+1}^T \mid C_t = i) = \Pr(\mathbf{X}^{(T)}, C_t = i).$$

Summation of this equation over i yields the second conclusion. $\qquad\square$

The second conclusion also follows immediately from the matrix expression for the likelihood and the definitions of $\boldsymbol{\alpha}_t$ and $\boldsymbol{\beta}_t$:

$$
\begin{aligned}
L_T &= \left(\boldsymbol{\delta}\mathbf{P}(x_1)\boldsymbol{\Gamma}\mathbf{P}(x_2)\ldots\boldsymbol{\Gamma}\mathbf{P}(x_t)\right)\left(\boldsymbol{\Gamma}\mathbf{P}(x_{t+1})\ldots\boldsymbol{\Gamma}\mathbf{P}(x_T)\mathbf{1}'\right) \\
&= \boldsymbol{\alpha}_t \boldsymbol{\beta}_t'.
\end{aligned}
$$

Note that we now have available T routes to the computation of the likelihood L_T, one for each possible value of t. But the route we have used so far (the case $t = T$, yielding $L_T = \boldsymbol{\alpha}_T \mathbf{1}'$) seems the most convenient, as it requires the computation of forward probabilities only, and only a single pass (forward) through the data.

In applying the EM algorithm to HMMs we shall also need the following two properties.

Proposition 5 *Firstly, for* $t = 1, \ldots, T$,

$$\Pr(C_t = j \mid \mathbf{X}^{(T)} = \mathbf{x}^{(T)}) = \alpha_t(j)\beta_t(j)/L_T; \qquad (4.10)$$

and secondly, for $t = 2, \ldots, T$,

$$\Pr(C_{t-1} = j, C_t = k \mid \mathbf{X}^{(T)} = \mathbf{x}^{(T)}) = \alpha_{t-1}(j)\,\gamma_{jk}\,p_k(x_t)\,\beta_t(k)/L_T. \qquad (4.11)$$

Proof. The first assertion follows immediately from (4.9) above. The second is an application of Equations (B.4) and (B.5) of Appendix B, and the proof proceeds as follows.

$$
\begin{aligned}
&\Pr(C_{t-1} = j, C_t = k \mid \mathbf{X}^{(T)} = \mathbf{x}^{(T)}) \\
&= \Pr(\mathbf{X}^{(T)}, C_{t-1} = j, C_t = k)/L_T \\
&= \Pr(\mathbf{X}^{(t-1)}, C_{t-1} = j)\,\Pr(C_t = k \mid C_{t-1} = j)\,\Pr(\mathbf{X}_t^T \mid C_t = k)/L_T \\
&\qquad\qquad \text{by (B.4)} \\
&= \alpha_{t-1}(j)\,\gamma_{jk}\left(\Pr(X_t \mid C_t = k)\,\Pr(\mathbf{X}_{t+1}^T \mid C_t = k)\right)/L_T \\
&\qquad\qquad \text{by (B.5)} \\
&= \alpha_{t-1}(j)\,\gamma_{jk}\,p_k(x_t)\,\beta_t(k)/L_T. \qquad\qquad \square
\end{aligned}
$$

4.2 The EM algorithm

Since the sequence of states occupied by the Markov-chain component of an HMM is not observed, a very natural approach to parameter estimation in HMMs is to treat those states as missing data and to employ the EM algorithm (Dempster, Laird and Rubin, 1977) in order to find maximum likelihood estimates of the parameters. Indeed the pioneering work of Leonard Baum and his co-authors (see Baum *et al.*, 1970; Baum, 1972; Welch, 2003) on what later were called HMMs was an important precursor of the work of Dempster *et al.*

4.2.1 EM in general

The EM algorithm is an iterative method for performing maximum likelihood estimation when some of the data are missing, and exploits the fact that the complete-data log-likelihood may be straightforward to maximize even if the likelihood of the observed data is not. By 'complete-data log-likelihood' (CDLL) we mean the log-likelihood of the parameters of interest θ, based on both the observed data and the missing data.

The algorithm may be described informally as follows (see e.g. Little and Rubin, 2002, pp. 166–168). Choose starting values for the parameters θ you wish to estimate. Then repeat the following steps.

- **E step** Compute the conditional expectations of the missing data given the observations and given the current estimate of θ. More precisely, compute the conditional expectations of *those functions of* the missing data that appear in the complete-data log-likelihood.

- **M step** Maximize, with respect to θ, the complete-data log-likelihood with the functions of the missing data replaced in it by their conditional expectations.

These two steps are repeated until some convergence criterion has been satisfied, e.g. until the resulting change in θ is less than some threshold. The resulting value of θ is then a stationary point of the likelihood of the observed data. In some cases, however, the stationary point reached can be a local (as opposed to global) maximum or a saddle point.

Little and Rubin (p. 168) stress the point that it is not (necessarily) the missing data themselves that are replaced in the CDLL by their conditional expectations, but those functions of the missing data that appear in the CDLL; they describe this as the 'key idea of EM'.

4.2.2 EM for HMMs

In the case of an HMM it is here convenient to represent the sequence of states $c_1, c_2, \ldots, c_T$ followed by the Markov chain by the zero-one random variables defined as follows:

$$u_j(t) = 1 \text{ if and only if } c_t = j, \quad (t = 1, 2, \ldots, T)$$

and

$$v_{jk}(t) = 1 \text{ if and only if } c_{t-1} = j \text{ and } c_t = k \quad (t = 2, 3, \ldots, T).$$

With this notation, the complete-data log-likelihood of an HMM — i.e. the log-likelihood of the observations $x_1, x_2, \ldots, x_T$ plus the missing data $c_1, c_2, \ldots, c_T$ — is given by

$$\log\left(\Pr(\mathbf{x}^{(T)}, \mathbf{c}^{(T)})\right) = \log\left(\delta_{c_1} \prod_{t=2}^{T} \gamma_{c_{t-1}, c_t} \prod_{t=1}^{T} p_{c_t}(x_t)\right)$$

$$= \log \delta_{c_1} + \sum_{t=2}^{T} \log \gamma_{c_{t-1}, c_t} + \sum_{t=1}^{T} \log p_{c_t}(x_t).$$

Hence the CDLL is

$$\log \left(\Pr(\mathbf{x}^{(T)}, \mathbf{c}^{(T)}) \right)$$

$$= \sum_{j=1}^{m} u_j(1) \log \delta_j + \sum_{j=1}^{m} \sum_{k=1}^{m} \left(\sum_{t=2}^{T} v_{jk}(t) \right) \log \gamma_{jk}$$

$$+ \sum_{j=1}^{m} \sum_{t=1}^{T} u_j(t) \log p_j(x_t) \quad (4.12)$$

$$= \text{term } 1 + \text{term } 2 + \text{term } 3.$$

Here δ is to be understood as the *initial* distribution of the Markov chain, the distribution of C_1, not necessarily the stationary distribution. Of course it is not reasonable to try to estimate the initial distribution from just one observation at time 1, especially as the state of the Markov chain itself is not observed. It is therefore interesting to see how the EM algorithm responds to this unreasonable request: see Section 4.2.4. We shall later (Section 4.2.5) make the additional assumption that the Markov chain is stationary, and not merely homogeneous; δ will then denote the stationary distribution implied by $\boldsymbol{\Gamma}$, and the question of estimating δ will fall away.

The EM algorithm for HMMs proceeds as follows.

- **E step** Replace all the quantities $v_{jk}(t)$ and $u_j(t)$ by their conditional expectations given the observations $\mathbf{x}^{(T)}$ (and given the current parameter estimates):

$$\hat{u}_j(t) = \Pr(C_t{=}j \mid \mathbf{x}^{(T)}) = \alpha_t(j)\beta_t(j)/L_T; \quad (4.13)$$

and

$$\hat{v}_{jk}(t) = \Pr(C_{t-1}{=}j, C_t{=}k \mid \mathbf{x}^{(T)}) = \alpha_{t-1}(j)\,\gamma_{jk}\,p_k(x_t)\,\beta_t(k)/L_T. \quad (4.14)$$

(See Section 4.1.3, Equations (4.10) and (4.11), for justification of the above equalities.) Note that in this context we need the forward probabilities as computed for an HMM that does *not* assume stationarity of the underlying Markov chain $\{C_t\}$; the backward probabilities are however not affected by the stationarity or otherwise of $\{C_t\}$.

- **M step** Having replaced $v_{jk}(t)$ and $u_j(t)$ by $\hat{v}_{jk}(t)$ and $\hat{u}_j(t)$, maximize the CDLL, expression (4.12), with respect to the three sets of parameters: the initial distribution δ, the transition probability matrix $\boldsymbol{\Gamma}$, and the parameters of the state-dependent distributions (e.g. $\lambda_1, \ldots, \lambda_m$ in the case of a simple Poisson–HMM).

Examination of (4.12) reveals that the M step splits neatly into three separate maximizations, since (of the parameters) term 1 depends only on the initial distribution δ, term 2 on the transition probability matrix

$\boldsymbol{\Gamma}$, and term 3 on the 'state-dependent parameters'. We must therefore maximize:

1. $\sum_{j=1}^{m} \hat{u}_j(1) \log \delta_j$ with respect to $\boldsymbol{\delta}$;

2. $\sum_{j=1}^{m} \sum_{k=1}^{m} \left(\sum_{t=2}^{T} \hat{v}_{jk}(t) \right) \log \gamma_{jk}$ with respect to $\boldsymbol{\Gamma}$; and

3. $\sum_{j=1}^{m} \sum_{t=1}^{T} \hat{u}_j(t) \log p_j(x_t)$ with respect to the state-dependent parameters. Notice here that the only parameters on which the term $\sum_{t=1}^{T} \hat{u}_j(t) \log p_j(x_t)$ depends are those of the jth state-dependent distribution, p_j; this further simplifies the problem.

The solution is as follows.

1. Set $\delta_j = \hat{u}_j(1)/\sum_{j=1}^{m} \hat{u}_j(1) = \hat{u}_j(1)$. (See Exercise 11 of Chapter 1 for justification.)

2. Set $\gamma_{jk} = f_{jk}/\sum_{k=1}^{m} f_{jk}$, where $f_{jk} = \sum_{t=2}^{T} \hat{v}_{jk}(t)$. (Apply the result of Exercise 11 of Chapter 1 to each row.)

3. The maximization of the third term may be easy or difficult, depending on the nature of the state-dependent distributions assumed. It is essentially the standard problem of maximum likelihood estimation for the distributions concerned. In the case of Poisson and normal distributions, closed-form solutions are available: see Section 4.2.3. In some other cases, e.g. the gamma distributions and the negative binomial, numerical maximization will be necessary to carry out this part of the M step.

From point 2 above, we see that it is not the quantities $\hat{v}_{jk}(t)$ themselves that are needed, but their sums f_{jk}. It is worth noting that the computation of the forward and backward probabilities is susceptible to under- or overflow error, as are the computation and summation of the quantities $\hat{v}_{jk}(t)$. In applying EM as described here, precautions (e.g. scaling) therefore have to be taken in order to prevent, or at least reduce the risk of, such error. Code for computing the logarithms of the forward and backward probabilities of a Poisson–HMM, and for computing MLEs via the EM algorithm, appears in A.2.2 and A.2.3.

4.2.3 M step for Poisson– and normal–HMMs

Here we give part 3 of the M step explicitly for the cases of Poisson and normal state-dependent distributions. The state-dependent part of the CDLL (term 3 of expression (4.12)) is

$$\sum_{j=1}^{m} \sum_{t=1}^{T} \hat{u}_j(t) \log p_j(x_t).$$

For a Poisson–HMM, $p_j(x) = e^{-\lambda_j}\lambda_j^x/x!$, so in that case term 3 is maximized by setting

$$0 = \sum_t \hat{u}_j(t)(-1 + x_t/\lambda_j);$$

that is, by

$$\hat{\lambda}_j = \sum_{t=1}^{T} \hat{u}_j(t)x_t \Big/ \sum_{t=1}^{T} \hat{u}_j(t).$$

For a normal–HMM the state-dependent density is of the form $p_j(x) = (2\pi\sigma_j^2)^{-1/2}\exp\left(-\frac{1}{2\sigma_j^2}(x - \mu_j)^2\right)$, and the maximizing values of the state-dependent parameters μ_j and σ_j^2 are

$$\hat{\mu}_j = \sum_{t=1}^{T} \hat{u}_j(t)x_t \Big/ \sum_{t=1}^{T} \hat{u}_j(t),$$

and

$$\hat{\sigma}_j^2 = \sum_{t=1}^{T} \hat{u}_j(t)(x_t - \hat{\mu}_j)^2 \Big/ \sum_{t=1}^{T} \hat{u}_j(t).$$

4.2.4 Starting from a specified state

What is sometimes done, however, notably by Leroux and Puterman (1992), is instead to condition the likelihood of the observations — i.e. the 'incomplete-data' likelihood — on the Markov chain starting in a particular state: that is, to assume that $\boldsymbol{\delta}$ is a unit vector $(0, \ldots, 0, 1, 0, \ldots, 0)$ rather than a vector whose components δ_j all require estimation. But, since it is known (see Levinson, Rabiner and Sondhi, 1983, p. 1055) that, at a maximum of the likelihood, $\boldsymbol{\delta}$ is one of the m unit vectors, maximizing the conditional likelihood of the observations over the m possible starting states is equivalent to maximizing the unconditional likelihood over $\boldsymbol{\delta}$; see also Exercise 1. Furthermore, it requires considerably less computational effort. If one wishes to use EM in order to fit an HMM in which the Markov chain is not assumed stationary, this does seem to be the most sensible approach. In our EM examples, however, we shall treat $\boldsymbol{\delta}$ as a vector of parameters requiring estimation, as it is instructive to see what emerges.

4.2.5 EM for the case in which the Markov chain is stationary

Now assume in addition that the underlying Markov chain is stationary, and not merely homogeneous. This is often a desirable assumption in time series applications. The initial distribution $\boldsymbol{\delta}$ is then such that $\boldsymbol{\delta} =$

$\delta\boldsymbol{\Gamma}$ and $\delta\mathbf{1}' = 1$, or equivalently

$$\boldsymbol{\delta} = \mathbf{1}(\mathbf{I}_m - \boldsymbol{\Gamma} + \mathbf{U})^{-1},$$

with $\mathbf{U}$ being a square matrix of ones. In this case, $\boldsymbol{\delta}$ is completely determined by the transition probabilities $\boldsymbol{\Gamma}$, and the question of estimating $\boldsymbol{\delta}$ falls away. However, the M step then gives rise to the following optimization problem: maximize, with respect to $\boldsymbol{\Gamma}$, the sum of terms 1 and 2 of expression (4.12), i.e. maximize

$$\sum_{j=1}^{m} \hat{u}_j(1) \log \delta_j + \sum_{j=1}^{m} \sum_{k=1}^{m} \left(\sum_{t=2}^{T} \hat{v}_{jk}(t) \right) \log \gamma_{jk}. \qquad (4.15)$$

Notice that here term 1 also depends on $\boldsymbol{\Gamma}$. Even in the case of only two states, analytic maximization would require the solution of a pair of quadratic equations in two variables, viz. two of the transition probabilities; see Exercise 3. Numerical solution is in general therefore needed for this part of the M step if stationarity is assumed. This is a slight disadvantage of the use of EM, as the stationary version of the models is important in a time series context.

4.3 Examples of EM applied to Poisson–HMMs

4.3.1 Earthquakes

We now present two- and three-state models fitted by the EM algorithm, as described above, to the earthquakes data. For the two-state model, the starting values of the off-diagonal transition probabilities are taken to be 0.1, and the starting value of $\boldsymbol{\delta}$, the initial distribution, is (0.5, 0.5). Since 19.36 is the sample mean, 10 and 30 are plausible starting values for the state-dependent means λ_1 and λ_2.

In the tables shown, 'iteration 0' refers to the starting values, and 'stationary model' to the parameter values and log-likelihood of the comparable stationary model fitted via nlm by direct numerical maximization.

Several features of Table 4.1 are worth noting. Firstly, the likelihood value of the stationary model is slightly lower than that fitted here by EM (i.e. $-l$ is higher). This is to be expected, as constraining the initial distribution $\boldsymbol{\delta}$ to be the stationary distribution can only decrease the maximal value of the likelihood. Secondly, the estimates of the transition probabilities and the state-dependent means are not identical for the two models, but close; this, too, is to be expected. Thirdly, although we know from Section 4.2.4 that $\boldsymbol{\delta}$ will approach a unit vector, it is noticeable just how quickly, starting from (0.5, 0.5), it approaches (1, 0).

Table 4.2 displays similar information for the corresponding three-state models. In this case as well, the starting values of the off-diagonal

Table 4.1 *Two-state model for earthquakes, fitted by EM.*

Iteration	γ_{12}	γ_{21}	λ_1	λ_2	δ_1	$-l$
0	0.100000	0.10000	10.000	30.000	0.50000	413.27542
1	0.138816	0.11622	13.742	24.169	0.99963	343.76023
2	0.115510	0.10079	14.090	24.061	1.00000	343.13618
30	0.071653	0.11895	15.419	26.014	1.00000	341.87871
50	0.071626	0.11903	15.421	26.018	1.00000	341.87870
convergence	0.071626	0.11903	15.421	26.018	1.00000	341.87870
stationary model	0.065961	0.12851	15.472	26.125	0.66082	342.31827

Table 4.2 *Three-state model for earthquakes, fitted by EM.*

Iteration	λ_1	λ_2	λ_3	δ_1	δ_2	$-l$
0	10.000	20.000	30.000	0.33333	0.33333	342.90781
1	11.699	19.030	29.741	0.92471	0.07487	332.12143
2	12.265	19.078	29.581	0.99588	0.00412	330.63689
30	13.134	19.713	29.710	1.00000	0.00000	328.52748
convergence	13.134	19.713	29.710	1.00000	0.00000	328.52748
stationary model	13.146	19.721	29.714	0.44364	0.40450	329.46028

transition probabilities are all taken to be 0.1 and the starting δ is uniform over the states.

We now present more fully the 'EM' and the stationary versions of the three-state model, which are only summarized in Table 4.2.

- Three-state model with initial distribution (1,0,0), fitted by EM:

$$\Gamma = \begin{pmatrix} 0.9393 & 0.0321 & 0.0286 \\ 0.0404 & 0.9064 & 0.0532 \\ 0.0000 & 0.1903 & 0.8097 \end{pmatrix},$$

$$\lambda = (13.134, 19.713, 29.710).$$

- Three-state model based on stationary Markov chain, fitted by direct numerical maximization:

$$\Gamma = \begin{pmatrix} 0.9546 & 0.0244 & 0.0209 \\ 0.0498 & 0.8994 & 0.0509 \\ 0.0000 & 0.1966 & 0.8034 \end{pmatrix},$$

$$\delta = (0.4436, 0.4045, 0.1519),$$

$$\text{and } \lambda = (13.146, 19.721, 29.714).$$

One of the many things that can be done by means of the **R** package
msm (Jackson *et al.*, 2003) is to fit HMMs with a range of state-dependent
distributions, including the Poisson. The default initial distribution used
by msm assigns a probability of 1 to state 1; the resulting models are
therefore directly comparable to the models which we have fitted by EM.
The two-state model fitted by msm corresponds closely to our model as
given in Table 4.1; it has $-l = 341.8787$, state-dependent means 15.420
and 26.017, and transition probability matrix

$$\begin{pmatrix} 0.9283 & 0.0717 \\ 0.1189 & 0.8811 \end{pmatrix}.$$

The three-state models do not correspond quite so closely. The 'best'
three-state model we have found by msm has $-l = 328.6208$ (cf. 328.5275),
state-dependent means 13.096, 19.708 and 29.904, and transition probability matrix

$$\begin{pmatrix} 0.9371 & 0.0296 & 0.0333 \\ 0.0374 & 0.9148 & 0.0479 \\ 0.0040 & 0.1925 & 0.8035 \end{pmatrix}.$$

4.3.2 Foetal movement counts

Leroux and Puterman (1992) used EM to fit (among other models)
Markov-dependent mixtures to a time series of unbounded counts, a se-
ries which has subsequently been analysed by Chib (1996), Robert and
Titterington (1998), Robert and Casella (1999, p. 432) and Scott (2002).
The series consists of counts of movements by a foetal lamb in 240 con-
secutive 5-second intervals, and was taken from Wittmann, Rurak and
Taylor (1984).

We present here a two-state HMM fitted by EM, and two HMMs fitted
by direct numerical maximization of the likelihood, and we compare
these with the models in Table 4 of Leroux and Puterman and Table 2 of
Robert and Titterington. Robert and Titterington omit any comparison
with the results of Leroux and Puterman, a comparison which might have
been informative: the estimates of λ_1 differ, and Leroux and Puterman's
likelihood value is somewhat higher, as are our values.

The models of Leroux and Puterman start with probability 1 from
one of the states; i.e. the initial distribution of the Markov chain is a
unit vector. But it is of course easy to fit such models by direct numerical maximization if one so wishes. In any code which takes the initial
distribution of the Markov chain to be the stationary distribution, one
merely replaces that stationary distribution by the relevant unit vector.
Our Table 4.3 presents (*inter alia*) three models we have fitted, the second of which is of this kind. The first we have fitted by direct numerical

Table 4.3 *Six two-state models fitted to the foetal movement counts time series.*

	DNM (stat.)	DNM (state 2)	EM (this work)	EM (L&P)	SEM	prior feedback
λ_1	3.1148	3.1007	3.1007	3.1006	2.93	2.84
λ_2	0.2564	0.2560	0.2560	0.2560	0.26	0.25
γ_{12}	0.3103	0.3083	0.3083	0.3083	0.28	0.32
γ_{21}	0.0113	0.0116	0.0116	0.0116	0.01	0.015
$-l^*$	–	150.7007	150.7007	150.70	–	–
$-l$	177.5188	177.4833	177.4833	(177.48)	(177.58)	(177.56)
$10^{78}L$	8.0269	8.3174	8.3174	(8.3235)	7.539	7.686

Key to notation and abbreviations:

DNM (stat.)	model starting from stationary distribution of the Markov chain, fitted by direct numerical maximization
DNM (state 2)	model starting from state 2 of the Markov chain, fitted by direct numerical maximization
EM (this work)	model fitted by EM, as described in Sections 4.2.2–4.2.3
EM (L&P)	model fitted by EM by Leroux and Puterman (1992)
SEM	model fitted by stochastic EM algorithm of Chib (1996); from Table 2 of Robert and Titterington (1998)
prior feedback	model fitted by maximum likelihood via 'prior feedback'; from Table 2 of Robert and Titterington (1998)
l^*	log-likelihood omitting constant terms
l	log-likelihood
L	likelihood
–	not needed; deliberately omitted

Figures appearing in brackets in the last three columns of this table have been deduced from figures published in the works cited, but do not themselves appear there; they are at best as accurate as the figures on which they are based. For instance, $10^{78}L = 8.3235$ is based on $-l^* = 150.70$. The figures 150.70, 7.539 and 7.686 in those three columns are exactly as published.

---

maximization of the likelihood based on the initial distribution being the stationary distribution, the second by direct numerical maximization of the likelihood starting from state 2, and the third by EM. In Table 4.3 they are compared to three models appearing in the published literature. The very close correspondence between the models we fitted by EM and by direct maximization (starting in state 2) is reassuring.

The results displayed in the table suggest that, if one wishes to fit models by maximum likelihood, then EM and direct numerical maximization are superior to the other methods considered. Robert and Titterington suggest that Chib's algorithm may not have converged, or may have converged to a different local optimum of the likelihood; their prior feedback results appear also to be suboptimal, however.

4.4 Discussion

As Bulla and Berzel (2008) point out, researchers and practitioners tend to use either EM or direct numerical maximization, but not both, to perform maximum likelihood estimation in HMMs, and each approach has its merits. However, one of the merits rightly claimed for EM in some generality turns out to be illusory in the context of HMMs. McLachlan and Krishnan (1997, p. 33) state of the EM algorithm in general — i.e. not specifically in the context of HMMs — that it is '[...] easy to program, since no evaluation of the likelihood nor its derivatives is involved.' If one applies EM as described above, one has to compute both the forward and the backward probabilities; on the other hand one needs only the forward probabilities to compute the likelihood, which can then be maximized numerically. In effect, EM does all that is needed to compute the likelihood via the forward probabilities, and then does more. Especially if one has available an optimization routine, such as nlm, optim or constrOptim in **R**, which does not demand the specification of derivatives, ease of programming seems in the present context to be more a characteristic of direct numerical maximization than of EM.

In our experience, it is a major advantage of direct numerical maximization without analytical derivatives that one can, with a minimum of programming effort, repeatedly modify a model in an interactive search for the best model. Often all that is needed is a small change to the code that evaluates the likelihood. It is also usually straightforward to replace one optimizer by another if an optimizer fails, or if one wishes to check in any way the output of a particular optimizer.

Note the experience reported by Altman and Petkau (2005). In their applications direct maximization of the likelihood produced the MLEs far more quickly than did the EM algorithm. See also Turner (2008), who provides a detailed study of direct maximization by the Levenberg–Marquardt algorithm. In two examples he finds that this algorithm is much faster (in the sense of CPU time) than is EM, and it is also clearly faster than optim, both with and without the provision of analytical derivatives. In our opinion the disadvantage of using analytical derivatives in exploratory modelling is the work involved in recoding those derivatives, and checking the code, when one alters a model. Of course for standard models such as the Poisson–HMM, which are likely to be used repeatedly, the advantage of having efficient code would make such labour worthwhile. Cappé *et al.* (2005, p. 358) provide a discussion of the relative merits of EM and direct maximization of the likelihood of an HMM by gradient-based methods.

Exercises

1.(a) Suppose $L_i > 0$ for $i = 1, 2, \ldots, m$. Maximize $L = \sum_{i=1}^m a_i L_i$ over $a_i \geq 0$, $\sum_{i=1}^m a_i = 1$.

 (b) Consider an HMM with initial distribution $\boldsymbol{\delta}$, and consider the likelihood as a function of $\boldsymbol{\delta}$. Show that, at a maximum of the likelihood, $\boldsymbol{\delta}$ is a unit vector.

2. Consider the example on pp. 186–187 of Visser, Raijmakers and Molenaar (2002). There a series of length 1000 is simulated from an HMM with states S_1 and S_2 and the three observation symbols 1, 2 and 3. The transition probability matrix is

$$\mathbf{A} = \begin{pmatrix} 0.9 & 0.1 \\ 0.3 & 0.7 \end{pmatrix},$$

the initial probabilities are $\boldsymbol{\pi} = (0.5, 0.5)$, and the state-dependent distribution in state i is row i of the matrix

$$\mathbf{B} = \begin{pmatrix} 0.7 & 0.0 & 0.3 \\ 0.0 & 0.4 & 0.6 \end{pmatrix}.$$

The parameters $\mathbf{A}$, $\mathbf{B}$ and $\boldsymbol{\pi}$ are then estimated by EM; the estimates of $\mathbf{A}$ and $\mathbf{B}$ are close to $\mathbf{A}$ and $\mathbf{B}$, but that of $\boldsymbol{\pi}$ is $(1, 0)$. This estimate for $\boldsymbol{\pi}$ is explained as follows: 'The reason for this is that the sequence of symbols that was generated actually starts with the symbol 1 which can only be produced from state S_1.'

Do you agree with the above statement? What if the probability of symbol 1 in state S_2 had been (say) 0.1 rather than 0.0?

3. Consider the fitting by EM of a two-state HMM based on a *stationary* Markov chain. In the M step, the sum of terms 1 and 2 must be maximized with respect to $\boldsymbol{\Gamma}$; see the expression labelled (4.15).

Write term 1 + term 2 as a function of γ_{12} and γ_{21}, the off-diagonal transition probabilities, and differentiate to find the equations satisfied by these probabilities at a stationary point. (You should find that the equations are quadratic in both γ_{12} and γ_{21}.)

4. Consider again the soap sales series introduced in Exercise 5 of Chapter 1.

Use the EM algorithm to fit Poisson–HMMs with two, three and four states to these data.

5. Let $\{X_t\}$ be an HMM on m states.

(a) Suppose the state-dependent distributions are binomial. More precisely, assume that

$$\Pr(X_t = x \mid C_t = j) = \binom{n_t}{x} p_j^x (1 - p_j)^{n_t - x}.$$

Find the value for p_j that will maximize the third term of Equation (4.12). (This is needed in order to carry out the M step of EM for a binomial–HMM.)

(b) Now suppose instead that the state-dependent distributions are exponential, with means $1/\lambda_j$. Find the value for λ_j that will maximize the third term of Equation (4.12).

6. Modify the code given in A.2 for Poisson–HMMs, in order to fit normal–, binomial–, and exponential–HMMs by EM.

CHAPTER 5

Forecasting, decoding and state prediction

Main results of this chapter:

conditional distributions   p. 77

$$\Pr\left(X_t = x \mid \mathbf{X}^{(-t)} = \mathbf{x}^{(-t)}\right) = \sum_i w_i(t) p_i(x)$$

forecast distributions   p. 79

$$\Pr(X_{T+h} = x \mid \mathbf{X}^{(T)} = \mathbf{x}^{(T)}) = \sum_i \xi_i(h) p_i(x)$$

state probabilities and local decoding   p. 81

$$\Pr(C_t = i \mid \mathbf{X}^{(T)} = \mathbf{x}^{(T)}) = \alpha_t(i)\beta_t(i)/L_T$$

global decoding maximize over $\mathbf{c}^{(T)}$:   p. 82

$$\Pr(\mathbf{C}^{(T)} = \mathbf{c}^{(T)} \mid \mathbf{X}^{(T)} = \mathbf{x}^{(T)})$$

state prediction   p. 86

$$\Pr(C_{T+h} = i \mid \mathbf{X}^{(T)} = \mathbf{x}^{(T)}) = \boldsymbol{\alpha}_T \boldsymbol{\Gamma}^h(, i)/L_T$$

Convenient expressions for conditional distributions and forecast distributions are available for HMMs. This makes it easy, for example, to check for outliers or to make interval forecasts. In this chapter, we first show (in Section 5.1) how to compute the conditional distribution of an observation under an HMM, i.e. the distribution of the observation at time t given the observations at all other times. In Section 5.2 we derive the forecast distribution of an HMM. Then, in Section 5.3, we demonstrate how, given the HMM and the observations, one can deduce information about the states occupied by the underlying Markov chain. Such inference is known as decoding. We continue to use the earthquakes series as our illustrative example. Our results are stated for the case of discrete observations X_t; if the observations are continuous, probabilities will need to be replaced by densities.

Note that in this chapter we do not assume stationarity of the Markov chain $\{C_t\}$, only homogeneity: here the row vector $\boldsymbol{\delta}$ denotes the *initial* distribution, that of C_1, and is not assumed to be the stationary distribution. Of course the results also hold in the special case in which the Markov chain is stationary, in which case $\boldsymbol{\delta}$ is both the initial and the stationary distribution.

5.1 Conditional distributions

We now derive a formula for the distribution of X_t conditioned on all the other observations of the HMM. We use the notation $\mathbf{X}^{(-t)}$ for the observations at all times other than t; that is, we define

$$\mathbf{X}^{(-t)} \equiv (X_1, \ldots, X_{t-1}, X_{t+1}, \ldots, X_T),$$

and similarly $\mathbf{x}^{(-t)}$.

Using the likelihood of an HMM as discussed in Section 2.3.2, and the definition of the forward and backward probabilities as in Section 4.1, it follows immediately, for $t = 2, 3, \ldots, T$, that

$$\Pr\left(X_t = x \mid \mathbf{X}^{(-t)} = \mathbf{x}^{(-t)}\right) = \frac{\boldsymbol{\delta}\mathbf{P}(x_1)\mathbf{B}_2 \cdots \mathbf{B}_{t-1}\boldsymbol{\Gamma}\mathbf{P}(x)\mathbf{B}_{t+1} \cdots \mathbf{B}_T\mathbf{1}'}{\boldsymbol{\delta}\mathbf{P}(x_1)\mathbf{B}_2 \cdots \mathbf{B}_{t-1}\boldsymbol{\Gamma}\mathbf{B}_{t+1} \cdots \mathbf{B}_T\mathbf{1}'}$$

$$\propto \boldsymbol{\alpha}_{t-1}\boldsymbol{\Gamma}\mathbf{P}(x)\boldsymbol{\beta}_t'. \tag{5.1}$$

Here, as before, $\boldsymbol{\alpha}_t = \boldsymbol{\delta}\mathbf{P}(x_1)\mathbf{B}_2 \cdots \mathbf{B}_t$, $\boldsymbol{\beta}_t' = \mathbf{B}_{t+1} \cdots \mathbf{B}_T\mathbf{1}'$ and $\beta_T = 1$; recall that $\mathbf{B}_t$ is defined as $\boldsymbol{\Gamma}\mathbf{P}(x_t)$.

The result for the case $t = 1$ is

$$\Pr\left(X_1 = x \mid \mathbf{X}^{(-1)} = \mathbf{x}^{(-1)}\right) = \frac{\boldsymbol{\delta}\mathbf{P}(x)\mathbf{B}_2 \cdots \mathbf{B}_T\mathbf{1}'}{\boldsymbol{\delta}\mathbf{I}\mathbf{B}_2 \cdots \mathbf{B}_T\mathbf{1}'}$$

$$\propto \boldsymbol{\delta}\mathbf{P}(x)\boldsymbol{\beta}_1'. \tag{5.2}$$

The above conditional distributions are ratios of two likelihoods of an HMM: the numerator is the likelihood of the observations except that the observation x_t is replaced by x, and the denominator (the reciprocal of the constant of proportionality) is the likelihood of the observations except that x_t is treated as missing.

We now show that these conditional probabilities can be expressed as mixtures of the m state-dependent probability distributions. In both Equations (5.1) and (5.2) the required conditional probability has the following form: a row vector multiplied by the $m \times m$ diagonal matrix $\mathbf{P}(x) = \text{diag}(p_1(x), \ldots, p_m(x))$, multiplied by a column vector. It follows, for $t = 1, 2, \ldots, T$, that

$$\Pr\left(X_1 = x \mid \mathbf{X}^{(-t)} = \mathbf{x}^{(-t)}\right) \propto \sum_{i=1}^{m} d_i(t)p_i(x),$$

where, in the case of (5.1), $d_i(t)$ is the product of the ith entry of the vector $\boldsymbol{\alpha}_{t-1}\boldsymbol{\Gamma}$ and the i th entry of the vector $\boldsymbol{\beta}_t$; and in the case of (5.2), it is the product of the i th entry of the vector $\boldsymbol{\delta}$ and the i th entry of the vector $\boldsymbol{\beta}_1$. Hence

$$\Pr\left(X_t = x \mid \mathbf{X}^{(-t)} = \mathbf{x}^{(-t)}\right) = \sum_{i=1}^{m} w_i(t)p_i(x),\qquad(5.3)$$

where the mixing probabilities $w_i(t) = d_i(t)/\sum_{j=1}^{m} d_j(t)$ are functions of the observations $\mathbf{x}^{(-t)}$ and of the model parameters. The $\mathbf{R}$ code for such conditional distributions is given in A.2.9.

In Figure 5.1 we present the full array of conditional distributions for the earthquakes data. It is clear that each of the conditional distributions has a different shape, and the shape may change sharply from one time-point to the next. In addition, it is striking that some of the observed counts, which in Figure 5.1 are marked as bold bars, are extreme relative to their conditional distributions. This observation suggests using the conditional distributions for outlier checking, which will be demonstrated in Section 6.2.

5.2 Forecast distributions

We turn now to another type of conditional distribution, the forecast distribution of an HMM. Specifically we derive two expressions for the conditional distribution of X_{T+h} given $\mathbf{X}^{(T)} = \mathbf{x}^{(T)}$; h is termed the forecast horizon. Again we shall focus on the discrete case; the formulae for the continuous case are the same but with the probability functions replaced by density functions.

For discrete-valued observations the forecast distribution $\Pr(X_{T+h} = x \mid \mathbf{X}^{(T)} = \mathbf{x}^{(T)})$ of an HMM is very similar to the conditional distribution $\Pr(X_t = x \mid \mathbf{X}^{(-t)} = \mathbf{x}^{(-t)})$ just discussed, and can be computed in essentially the same way, as a ratio of likelihoods:

$$\begin{aligned}
\Pr(X_{T+h} = x \mid \mathbf{X}^{(T)} = \mathbf{x}^{(T)}) &= \frac{\Pr(\mathbf{X}^{(T)} = \mathbf{x}^{(T)}, X_{T+h} = x)}{\Pr(\mathbf{X}^{(T)} = \mathbf{x}^{(T)})} \\
&= \frac{\boldsymbol{\delta}\mathbf{P}(x_1)\mathbf{B}_2\mathbf{B}_3 \cdots \mathbf{B}_T\boldsymbol{\Gamma}^h\mathbf{P}(x)\mathbf{1}'}{\boldsymbol{\delta}\mathbf{P}(x_1)\mathbf{B}_2\mathbf{B}_3 \cdots \mathbf{B}_T\mathbf{1}'} \\
&= \frac{\boldsymbol{\alpha}_T\boldsymbol{\Gamma}^h\mathbf{P}(x)\mathbf{1}'}{\boldsymbol{\alpha}_T\mathbf{1}'}.
\end{aligned}$$

Writing $\boldsymbol{\phi}_T = \boldsymbol{\alpha}_T/\boldsymbol{\alpha}_T\mathbf{1}'$ (see Section 3.2), we have

$$\Pr(X_{T+h} = x \mid \mathbf{X}^{(T)} = \mathbf{x}^{(T)}) = \boldsymbol{\phi}_T\boldsymbol{\Gamma}^h\mathbf{P}(x)\mathbf{1}'.\qquad(5.4)$$

Expressions for joint distributions of several forecasts can be derived along the same lines. (See Exercise 5.)

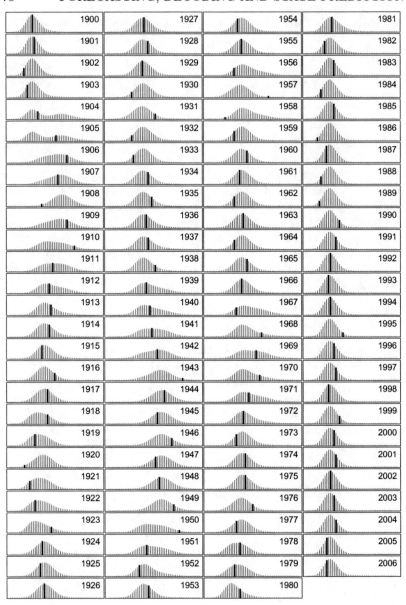

Figure 5.1 *Earthquakes data, three-state HMM: conditional distribution of the number of earthquakes in each year, given all the other observations. The bold bar corresponds to the actual number observed in that year.*

Table 5.1 *Earthquakes data, three-state Poisson–HMM: forecasts.*

year:	2007	2008	2009	2016	2026	2036
horizon:	1	2	5	10	20	30
forecast mode:	13	13	13	13	14	14
forecast median:	12.7	12.9	13.1	14.4	15.6	16.2
forecast mean:	13.7	14.1	14.5	16.4	17.5	18.0
nominal 90% forecast interval:	[8,21]	[8,23]	[8,25]	[8,30]	[8,32]	[9,32]
exact coverage:	0.908	0.907	0.907	0.918	0.932	0.910

The forecast distribution can therefore be written as a mixture of the state-dependent probability distributions:

$$\Pr(X_{T+h} = x \mid \mathbf{X}^{(T)} = \mathbf{x}^{(T)}) = \sum_{i=1}^{m} \xi_i(h) p_i(x), \qquad (5.5)$$

where the weight $\xi_i(h)$ is the ith entry of the vector $\boldsymbol{\phi}_T \boldsymbol{\Gamma}^h$. The **R** code for forecasts is given in A.2.8.

Since the entire probability distribution of the forecast is known, it is possible to make interval forecasts, and not only point forecasts. This is illustrated in Table 5.1, which lists statistics of some forecast distributions for the earthquake series fitted with a three-state Poisson HMM.

As the forecast horizon h increases, the forecast distribution converges to the marginal distribution of the stationary HMM, i.e.

$$\lim_{h \to \infty} \Pr(X_{T+h} = x \mid \mathbf{X}^{(T)} = \mathbf{x}^{(T)}) = \lim_{h \to \infty} \boldsymbol{\phi}_T \boldsymbol{\Gamma}^h \mathbf{P}(x) \mathbf{1}' = \boldsymbol{\delta}^* \mathbf{P}(x) \mathbf{1}',$$

where here we temporarily use $\boldsymbol{\delta}^*$ to denote the stationary distribution of the Markov chain (in order to distinguish it from $\boldsymbol{\delta}$, the initial distribution). The limit follows from the observation that, for any nonnegative (row) vector $\boldsymbol{\eta}$ whose entries add to 1, the vector $\boldsymbol{\eta} \boldsymbol{\Gamma}^h$ approaches the stationary distribution of the Markov chain as $h \to \infty$, provided the Markov chain satisfies the usual regularity conditions of irreducibility and aperiodicity; see e.g. Feller (1968, p. 394). Sometimes the forecast distribution approaches its limiting distribution only slowly; see Figure 5.2, which displays six of the forecast distributions for the earthquakes series, compared with the limiting distribution. In other cases the approach can be relatively fast; for a case in point, consider the three-state model for the soap sales series introduced in Exercise 5 of Chapter 1. The rate of approach is determined by the size of the largest eigenvalue other than 1 of the t.p.m. $\boldsymbol{\Gamma}$.

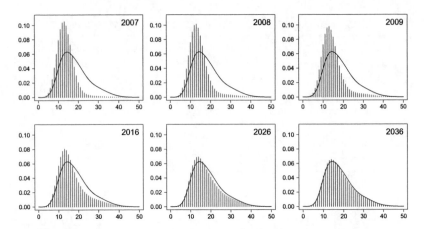

Figure 5.2 *Earthquakes data, three-state Poisson–HMM: forecast distributions for 1 to 30 years ahead, compared to limiting distribution, which is shown as a continuous line.*

5.3 Decoding

In speech recognition and other applications — see e.g. Fredkin and Rice (1992), or Guttorp (1995, p. 101) — it is of interest to determine the states of the Markov chain that are most likely (under the fitted model) to have given rise to the observation sequence. In the context of speech recognition this is known as the decoding problem: see Juang and Rabiner (1991). More specifically, 'local decoding' of the state at time t refers to the determination of that state which is most likely at that time. In contrast, 'global decoding' refers to the determination of the most likely sequence of states. These two are described in the next two sections.

5.3.1 State probabilities and local decoding

Consider again the vectors of forward and backward probabilities, α_t and β_t, as discussed in Section 4.1. For the derivation of the most likely state of the Markov chain at time t, we shall use the following result, which appears there as Equation (4.9):

$$\alpha_t(i)\beta_t(i) = \Pr(\mathbf{X}^{(T)} = \mathbf{x}^{(T)}, C_t = i).$$

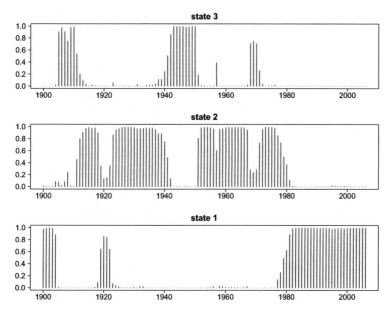

Figure 5.3 *Earthquakes data: state probabilities for fitted three-state HMM.*

Hence the conditional distribution of C_t given the observations can be obtained, for $i = 1, 2, \ldots, m$, as

$$
\begin{aligned}
\Pr(C_t = i \mid \mathbf{X}^{(T)} = \mathbf{x}^{(T)}) &= \frac{\Pr(C_t = i, \mathbf{X}^{(T)} = \mathbf{x}^{(T)})}{\Pr(\mathbf{X}^{(T)} = \mathbf{x}^{(T)})} \\
&= \frac{\alpha_t(i)\beta_t(i)}{L_T}.
\end{aligned}
\tag{5.6}
$$

Here L_T can be computed by the scaling method described in Section 3.2. Scaling is also necessary in order to prevent numerical underflow in the evaluation of the product $\alpha_t(i)\beta_t(i)$. The **R** code given in A.2.5 implements one method of doing this.

For each time $t \in \{1, \ldots, T\}$ one can therefore determine the distribution of the state C_t, given the observations $\mathbf{x}^{(T)}$, which for m states is a discrete probability distribution with support $\{1, \ldots, m\}$. In Figures 5.3 and 5.4 we display the state probabilities for the earthquakes series, based on the fitted three- and four-state Poisson–HMM models.

For each $t \in \{1, \ldots, T\}$ the most probable state i_t^*, given the observations, is defined as

$$
i_t^* = \operatorname*{argmax}_{i=1,\ldots,m} \Pr(C_t = i \mid \mathbf{X}^{(T)} = \mathbf{x}^{(T)}).
\tag{5.7}
$$

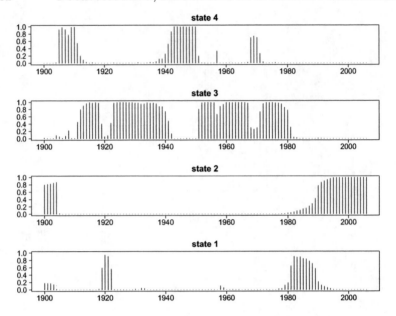

Figure 5.4 *Earthquakes data: state probabilities for fitted four-state HMM.*

This approach determines the most likely state separately for each t by maximizing the conditional probability $\Pr(C_t = i \mid \mathbf{X}^{(T)} = \mathbf{x}^{(T)})$ and is therefore called **local decoding**. In Figure 5.5 we display the results of applying local decoding to the earthquakes series, for the fitted three- and four-state models. The relevant **R** code is given in A.2.5 and A.2.6.

5.3.2 Global decoding

In many applications, e.g. speech recognition, one is not so much inter-ested in the most likely state for each separate time t — as provided by local decoding — as in the most likely *sequence* of (hidden) states. Instead of maximizing $\Pr(C_t = i \mid \mathbf{X}^{(T)} = \mathbf{x}^{(T)})$ over i for each t, one seeks that sequence of states $c_1, c_2, \ldots, c_T$ which maximizes the condi-tional probability

$$\Pr(\mathbf{C}^{(T)} = \mathbf{c}^{(T)} \mid \mathbf{X}^{(T)} = \mathbf{x}^{(T)}); \qquad (5.8)$$

or equivalently, and more conveniently, the joint probability:

$$\Pr(\mathbf{C}^{(T)}, \mathbf{X}^{(T)}) \;=\; \delta_{c_1} \prod_{t=2}^{T} \gamma_{c_{t-1}, c_t} \prod_{t=1}^{T} p_{c_t}(x_t).$$

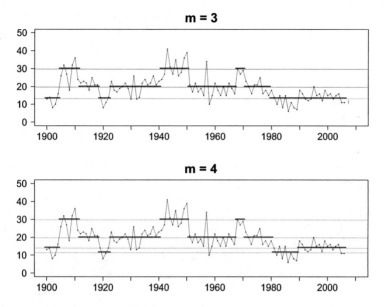

Figure 5.5 *Earthquakes data: local decoding according to three- and four-state HMMs. The horizontal lines indicate the state-dependent means.*

This is a subtly different maximization problem from that of local decoding, and is called **global decoding**. The results of local and global decoding are often very similar but not identical.

Maximizing (5.8) over all possible state sequences $c_1, c_2, \ldots, c_T$ by brute force would involve m^T function evaluations, which is clearly not feasible except for very small T. Fortunately one can use instead an efficient dynamic programming algorithm to determine the most likely sequence of states. The Viterbi algorithm (Viterbi, 1967; Forney, 1973) is such an algorithm.

We begin by defining

$$\xi_{1i} = \Pr(C_1 = i, X_1 = x_1) = \delta_i \, p_i(x_1),$$

and, for $t = 2, 3, \ldots, T$,

$$\xi_{ti} = \max_{c_1, c_2, \ldots, c_{t-1}} \Pr(\mathbf{C}^{(t-1)} = \mathbf{c}^{(t-1)}, C_t = i, \mathbf{X}^{(T)} = \mathbf{x}^{(T)}).$$

It can then be shown (see Exercise 1) that the probabilities ξ_{tj} satisfy the following recursion, for $t = 2, 3, \ldots, T$ and $i = 1, 2, \ldots, m$:

$$\xi_{tj} = \big(\max_i (\xi_{t-1,i} \, \gamma_{ij}) \big) p_j(x_t). \tag{5.9}$$

This provides an efficient means of computing the $T \times m$ matrix of values ξ_{tj}, as the computational effort is linear in T. The required maximizing sequence of states $i_1, i_2, \ldots, i_T$ can then be determined recursively from

$$i_T = \underset{i=1,\ldots,m}{\operatorname{argmax}} \, \xi_{Ti} \qquad (5.10)$$

and, for $t = T - 1, T - 2, \ldots, 1$, from

$$i_t = \underset{i=1,\ldots,m}{\operatorname{argmax}} \, (\xi_{ti} \, \gamma_{i,i_{t+1}}). \qquad (5.11)$$

Note that, since the quantity to be maximized in global decoding is simply a product of probabilities (as opposed to a sum of such products), one can choose to maximize its logarithm, in order to prevent numerical underflow; the Viterbi algorithm can easily be rewritten in terms of the logarithms of the probabilities. Alternatively a scaling similar to that used in the likelihood computation can be employed: in that case one scales each of the T rows of the matrix $\{\xi_{ti}\}$ to have row sum 1. The Viterbi algorithm is applicable to both stationary and nonstationary underlying Markov chains; there is no necessity to assume that the initial distribution δ is the stationary distribution. For the relevant **R** code, see A.2.4.

Figure 5.6 displays, for the fitted three- and four-state models for the earthquakes series, the paths obtained by the Viterbi algorithm. The paths are very similar to those obtained by local decoding: compare with Figure 5.5. But they do differ. In the case of the three-state model, the years 1911, 1941 and 1980 differ. In the case of the four-state model, 1911 and 1941 differ. Notice also the nature of the difference between the upper and lower panels of Figure 5.6: allowing for a fourth state has the effect of splitting one of the states of the three-state model, that with the lowest mean. When four states are allowed, the 'Viterbi path' moves along the lowest state in the years 1919–1922 and 1981–1989 only.

Global decoding is the main objective in many applications, especially when there are substantive interpretations for the states. It is therefore of interest to investigate the performance of global decoding in identifying the correct states. This can be done by simulating a series from an HMM, applying the algorithm in order to decode the simulated observations, and then comparing the Viterbi path with the (known) series of simulated states. We present here an example of such a comparison, based on a simulated sequence of length 100 000 from the three-state (stationary) model for the earthquakes given on p. 51.

The 3×3 table displayed below with its marginals gives the simulated joint distribution of the true state i (rows) and the Viterbi estimate j of the state (columns). The row totals are close to $(0.444, 0.405, 0.152)$, the stationary distribution of the model; this provides a partial check of the

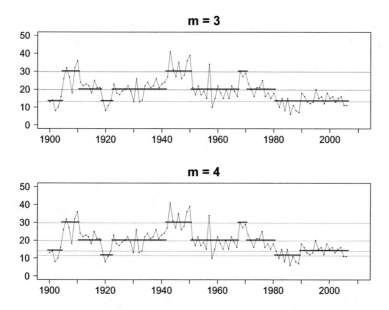

Figure 5.6 *Earthquakes: global decoding according to three- and four-state HMMs. The horizontal lines indicate the state-dependent means.*

simulation. The column totals (the distribution of the state as inferred by Viterbi) are also close to this stationary distribution.

	$j = 1$	2	3	
$i = 1$	0.431	0.017	0.000	0.448
2	0.018	0.366	0.013	0.398
3	0.000	0.020	0.134	0.154
	0.449	0.403	0.147	1.000

From this table one may conclude, for instance, that the estimated probability that the state inferred is 2, if the true state is 1, is $0.017/0.448 = 0.038$. More generally, the left-hand table below gives $\Pr(\text{inferred state} = j \mid \text{true state} = i)$ and the right-hand table gives $\Pr(\text{true state} = i \mid \text{inferred state} = j)$.

	$j = 1$	2	3		$j = 1$	2	3
$i = 1$	0.961	0.038	0.000	$i = 1$	0.958	0.043	0.001
2	0.046	0.921	0.033	2	0.041	0.908	0.088
3	0.002	0.130	0.868	3	0.001	0.050	0.910

Ideally all the diagonal elements of the above two tables would be 1; here

Table 5.2 *Earthquakes data. State prediction using a three-state Poisson–HMM: the probability that the Markov chain will be in a given state in the specified year.*

year	2007	2008	2009	2016	2026	2036
state=1	0.951	0.909	0.871	0.674	0.538	0.482
2	0.028	0.053	0.077	0.220	0.328	0.373
3	0.021	0.038	0.052	0.107	0.134	0.145

they range from 0.868 to 0.961. Such a simulation exercise quantifies the expected accuracy of the Viterbi path and is therefore particularly recommended in applications in which the interpretation of that path is an important objective of the analysis.

5.4 State prediction

In Section 5.3.1 we derived an expression for the conditional distribution of the state C_t, for $t = 1, 2, \ldots, T$, given the observations $\mathbf{x}^{(T)}$. In so doing we considered only present or past states. However, it is also possible to provide the conditional distribution of the state C_t for $t > T$, i.e to perform 'state prediction'.

Given the observations $x_1, \ldots, x_T$, the following set of statements can be made about future, present and past states (respectively):

$$
L_T \Pr(C_t = i \mid \mathbf{X}^{(T)} = \mathbf{x}^{(T)})
$$
$$
= \begin{cases} \boldsymbol{\alpha}_T \boldsymbol{\Gamma}^{t-T}(,i) & \text{for } t > T & \text{state prediction} \\ \alpha_T(i) & \text{for } t = T & \text{filtering} \\ \alpha_t(i)\beta_t(i) & \text{for } 1 \leq t < T & \text{smoothing,} \end{cases}
$$

where $\boldsymbol{\Gamma}^{t-T}(,i)$ denotes the ith column of the matrix $\boldsymbol{\Gamma}^{t-T}$. The 'filtering' and 'smoothing' parts (for present or past states) are identical to the state probabilities as described in Section 5.3.1, and indeed could here be combined, since $\beta_T(i) = 1$ for all i. The 'state prediction' part is simply a generalization to $t > T$, the future, and can be restated as follows (see Exercise 6); for $i = 1, 2, \ldots, m$,

$$
\Pr(C_{T+h} = i \mid \mathbf{X}^{(T)} = \mathbf{x}^{(T)}) = \boldsymbol{\alpha}_T \boldsymbol{\Gamma}^h(,i)/L_T = \boldsymbol{\phi}_T \boldsymbol{\Gamma}^h(,i), \quad (5.12)
$$

with $\boldsymbol{\phi}_T = \boldsymbol{\alpha}_T/\boldsymbol{\alpha}_T \mathbf{1}'$. Note that, as $h \to \infty$, $\boldsymbol{\phi}_T \boldsymbol{\Gamma}^h$ approaches the stationary distribution of the Markov chain.

Table 5.2 gives, for a range of years, the state predictions based on the three-state model for the earthquake series. The **R** code for state prediction is given in A.2.7.

Exercises

1. Prove the recursion (5.9):

$$\xi_{tj} = \left(\max_i (\xi_{t-1,i}\, \gamma_{ij}) \right) p_j(x_t).$$

2. Apply local and global decoding to a three-state model for the soap sales series introduced in Exercise 5 of Chapter 1, and compare the results to see how much the conclusions differ.

3. Compute the h-step-ahead state predictions for the soap sales series, for $h = 1$ to 5. How close are these distributions to the stationary distribution of the Markov chain?

4.(a) Using the same sequence of random numbers in each case, generate sequences of length 1000 from the Poisson–HMMs with

$$\Gamma = \begin{pmatrix} 0.8 & 0.1 & 0.1 \\ 0.1 & 0.8 & 0.1 \\ 0.1 & 0.1 & 0.8 \end{pmatrix},$$

and: (i) $\lambda = (10, 20, 30)$; and (ii) $\lambda = (15, 20, 25)$. Keep a record of the sequence of states, which should be the same in (i) and (ii).

(b) Use the Viterbi algorithm to infer the most likely sequence of states in each case, and compare these two sequences to the 'true' underlying sequence, i.e. the generated one.

(c) What conclusions do you draw about the accuracy of the Viterbi algorithm?

5. Bivariate forecast distributions for HMMs

(a) Find the joint distribution of X_{T+1} and X_{T+2}, given $\mathbf{X}^{(T)}$, in as simple a form as you can.

(b) For the earthquakes data, find $\Pr(X_{T+1} \leq 10, X_{T+2} \leq 10 \mid \mathbf{X}^{(T)})$.

6. Prove Equation (5.12):

$$\Pr(C_{T+h} = i \mid \mathbf{X}^{(T)} = \mathbf{x}^{(T)}) = \alpha_T \Gamma^h(, i)/L_T = \phi_T \Gamma^h(, i).$$

CHAPTER 6

Model selection and checking

In the basic HMM with m states, increasing m always improves the fit of the model (as judged by the likelihood). But along with the improvement comes a quadratic increase in the number of parameters, and the improvement in fit has to be traded off against this increase. A criterion for model selection is therefore needed.

In some cases, it is sensible to reduce the number of parameters by making assumptions on the state-dependent distributions or on the t.p.m. of the Markov chain. For an example of the former, see p. 175, where, in order to model a series of categorical observations with 16 circular categories, von Mises distributions (with two parameters) are used as the state-dependent distributions. For an example of the latter, see Section 13.3, where we describe discrete state-space stochastic volatility models which are m-state HMMs with only three or four parameters. Notice that in this case the number of parameters does not increase at all with increasing m.

In this chapter we give a brief account of model selection in HMMs (Section 6.1), and then describe the use of pseudo-residuals in order to check for deficiencies in the selected model (Section 6.2).

6.1 Model selection by AIC and BIC

A problem which arises naturally when one uses hidden Markov (or other) models is that of selecting an appropriate model, e.g. of choosing the appropriate number of states m, sometimes described as the 'order' of the HMM, or of choosing between competing state-dependent distributions such as Poisson and negative binomial. Although the question of order estimation for an HMM is neither trivial nor settled (see Cappé et al., 2005, Chapter 15), we need some criterion for model comparison. The material outlined below is based on Zucchini (2000), which gives an introductory account of model selection.

Assume that the observations $x_1, \ldots, x_T$ were generated by the unknown 'true' or 'operating' model f and that one fits models from two different approximating families, $\{g_1 \in G_1\}$ and $\{g_2 \in G_2\}$. The goal of model selection is to identify the model which is in some sense the best.

There exist at least two approaches to model selection. In the frequentist approach one selects the family estimated to be closest to the operating model. For that purpose one defines a discrepancy (a measure of 'lack of fit') between the operating and the fitted models, $\Delta(f, \hat{g}_1)$ and $\Delta(f, \hat{g}_2)$. These discrepancies depend on the operating model f, which is unknown, and so it is not possible to determine which of the two discrepancies is smaller, i.e. which model should be selected. Instead one bases selection on estimators of the expected discrepancies, namely $\widehat{E}_f(\Delta(f, \hat{g}_1))$ and $\widehat{E}_f(\Delta(f, \hat{g}_2))$, which are referred to as model selection criteria. By choosing the Kullback–Leibler discrepancy, and under the conditions listed in Appendix A of Linhart and Zucchini (1986), the model selection criterion simplifies to the Akaike information criterion (AIC):

$$\text{AIC} = -2 \log L + 2p,$$

where $\log L$ is the log-likelihood of the fitted model and p denotes the number of parameters of the model. The first term is a measure of fit, and decreases with increasing number of states m. The second term is a penalty term, and increases with increasing m.

The Bayesian approach to model selection is to select the family which is estimated to be most likely to be true. In a first step, before considering the observations, one specifies the priors, i.e. the probabilities $\Pr(f \in G_1)$ and $\Pr(f \in G_2)$ that f stems from the approximating family. In a second step one computes and compares the posteriors, i.e. the probabilities that f belongs to the approximating family, given the observations, $\Pr(f \in G_1 \mid \mathbf{x}^{(T)})$ and $\Pr(f \in G_2 \mid \mathbf{x}^{(T)})$. Under certain conditions (see e.g. Wasserman (2000)), this approach results in the Bayesian information criterion (BIC) which differs from AIC in the penalty term:

$$\text{BIC} = -2 \log L + p \log T,$$

where $\log L$ and p are as for AIC, and T is the number of observations. Compared to AIC, the penalty term of BIC has more weight for $T > e^2$, which holds in most applications. Thus the BIC often favours models with fewer parameters than does the AIC.

For the earthquakes series, AIC and BIC both select three states: see Figure 6.1, which plots AIC and BIC against the number of states m of the HMM. The values of the two criteria are provided in Table 6.1. These values are also compared to those of the independent mixture models of Section 1.2.4. Although the HMMs demand more parameters than the comparable independent mixtures, the resulting values of AIC and BIC are lower than those obtained for the independent mixtures.

Several comments arise from Table 6.1. Firstly, given the serial dependence manifested in Figure 2.1, it is not surprising that the independent mixture models do not perform well relative to the HMMs. Secondly, although it is perhaps obvious *a priori* that one should not even try to

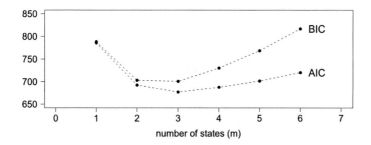

Figure 6.1 *Earthquakes series: model selection criteria AIC and BIC.*

Table 6.1 *Earthquakes data: comparison of (stationary) hidden Markov and independent mixture models by AIC and BIC.*

model	k	$-\log L$	AIC	BIC
'1-state HM'	1	391.9189	785.8	788.5
2-state HM	4	342.3183	692.6	703.3
3-state HM	9	329.4603	**676.9**	**701.0**
4-state HM	16	327.8316	687.7	730.4
5-state HM	25	325.9000	701.8	768.6
6-state HM	36	324.2270	720.5	816.7
indep. mixture (2)	3	360.3690	726.7	734.8
indep. mixture (3)	5	356.8489	723.7	737.1
indep. mixture (4)	7	356.7337	727.5	746.2

fit a model with as many as 25 or 36 parameters to 107 observations, and dependent observations at that, it is interesting to explore the likelihood functions in the case of HMMs with five and six states. The likelihood appears to be highly multimodal in these cases, and it is easy to find several local maxima by using different starting values. A strategy that seems to succeed in these cases is to start all the off-diagonal transition probabilities at small values (such as 0.01) and to space out the state-dependent means over a range somewhat less than the range of the observations.

According to both AIC and BIC, the model with three states is the most appropriate. But more generally the model selected may depend on the selection criterion adopted. The selected model is displayed on p. 51, and the state-dependent distributions, together with the resulting marginal, are displayed in Figure 3.1 on p. 51.

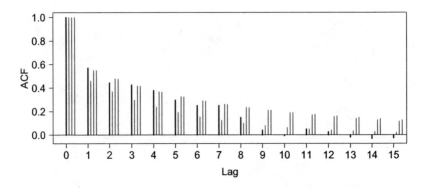

Figure 6.2 *Earthquakes data: sample ACF and ACF of three models. The bold bars on the left represent the sample ACF, and the other bars those of the HMMs with (from left to right) two, three and four states.*

It is also useful to compare the autocorrelation functions of the HMMs with two to four states with the sample ACF. The ACFs of the models can be found by using the results of Exercise 4 of Chapter 2, and appeared on p. 52. In tabular form the ACFs are:

k:	1	2	3	4	5	6	7	8
observations	0.570	0.444	0.426	0.379	0.297	0.251	0.251	0.149
2-state model	0.460	0.371	0.299	0.241	0.194	0.156	0.126	0.101
3-state model	0.551	0.479	0.419	0.370	0.328	0.292	0.261	0.235
4-state model	0.550	0.477	0.416	0.366	0.324	0.289	0.259	0.234

In Figure 6.2 the sample ACF is juxtaposed to those of the models with two, three and four states. It is clear that the autocorrelation functions of the models with three and four states correspond well to the sample ACF up to about lag 7. However, one can apply more systematic diagnostics, as will now be shown.

6.2 Model checking with pseudo-residuals

Even when one has selected what is by some criterion the 'best' model, there still remains the problem of deciding whether the model is indeed adequate; one needs tools to assess the general goodness of fit of the model, and to identify outliers relative to the model. In the simpler context of normal-theory regression models (for instance), the role of residuals as a tool for model checking is very well established. In this section

we describe quantities we call pseudo-residuals which are intended to ful-
fil this role much more generally, and which are useful in the context of
HMMs. We consider two versions of these pseudo-residuals (in Sections
6.2.2 and 6.2.3); both rely on being able to perform likelihood computations routinely, which is certainly the case for HMMs. A detailed account,
in German, of the construction and application of pseudo-residuals is
provided by Stadie (2002). See also Zucchini and MacDonald (1999).

6.2.1 Introducing pseudo-residuals

To motivate pseudo-residuals we need the following simple result. Let
X be a random variable with continuous distribution function F. Then
$U \equiv F(X)$ is uniformly distributed on the unit interval, which we write

$$U \sim U(0, 1).$$

The proof is left to the reader as Exercise 1.

The **uniform pseudo-residual** of an observation x_t from a continuous random variable X_t is defined as the probability, under the fitted
model, of obtaining an observation less than or equal to x_t:

$$u_t = \Pr(X_t \leq x_t) = F_{X_t}(x_t).$$

That is, u_t is the observation x_t transformed by its distribution function
under the model. If the model is correct, this type of pseudo-residual
is distributed U(0,1), with residuals for extreme observations close to
0 or 1. With the help of these uniform pseudo-residuals, observations
from different distributions can be compared. If we have observations
$x_1, \ldots, x_T$ and a model $X_t \sim F_t$, for $t = 1, \ldots, T$ (i.e. each x_t has its
own distribution function F_t), then the x_t-values cannot be compared
directly. However, the pseudo-residuals u_t are identically U(0,1) (if the
model is true), and can sensibly be compared. If a histogram or quantile-
quantile plot ('qq-plot') of the uniform pseudo-residuals u_t casts doubt
on the conclusion that they are U(0,1), one can deduce that the model
is not valid.

Although the uniform pseudo-residual is useful in this way, it has a
drawback if used for outlier identification. For example, if one considers
the values lying close to 0 or 1 on an index plot, it is hard to see whether
a value is very unlikely or not. A value of 0.999, for instance, is difficult
to distinguish from a value of 0.97, and an index plot is almost useless
for a quick visual analysis.

This deficiency of uniform pseudo-residuals can, however, easily be
remedied by using the following result. Let Φ be the distribution function of the standard normal distribution and X a random variable with
distribution function F. Then $Z \equiv \Phi^{-1}(F(X))$ is distributed standard

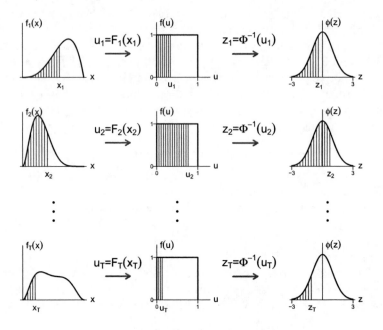

Figure 6.3 *Construction of normal pseudo-residuals in the continuous case.*

normal. (For the proof we refer again to Exercise 1.) We now define the
normal pseudo-residual as

$$z_t = \Phi^{-1}(u_t) = \Phi^{-1}(F_{X_t}(x_t)).$$

If the fitted model is valid, these normal pseudo-residuals are distributed
standard normal, with the value of the residual equal to 0 when the observation coincides with the median. Note that, by their definition, normal
pseudo-residuals measure the deviation from the median, and not from
the expectation. The construction of normal pseudo-residuals is illustrated in Figure 6.3. If the observations $x_1, \ldots, x_T$ were indeed generated
by the model $X_t \sim F_t$, the normal pseudo-residuals z_t would follow a
standard normal distribution. One can therefore check the model either
by visually analysing the histogram or qq-plot of the normal pseudo-
residuals, or by performing tests for normality.

This normal version of pseudo-residuals has the advantage that the
absolute value of the residual increases with increasing deviation from
the median and that extreme observations can be identified more easily
on a normal scale. This becomes obvious if one compares index plots of
uniform and normal pseudo-residuals.

Note that the theory of pseudo-residuals as outlined so far can be

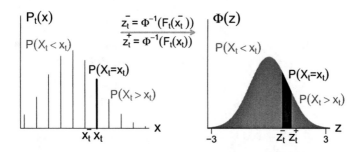

Figure 6.4 *Construction of normal pseudo-residuals in the discrete case.*

applied to continuous distributions only. In the case of discrete observations, the pseudo-residuals can, however, be modified to allow for the
discreteness. The pseudo-residuals are no longer defined as points, but
as intervals. Thus, for a discrete random variable X_t with distribution
function F_{X_t} we define the uniform pseudo-residual segments as

$$[u_t^-; u_t^+] = [F_{X_t}(x_t^-); F_{X_t}(x_t)], \qquad (6.1)$$

with x_t^- denoting the greatest realization possible that is strictly less
than x_t, and we define the normal pseudo-residual segments as

$$[z_t^-; z_t^+] = [\Phi^{-1}(u_t^-); \Phi^{-1}(u_t^+)] = [\Phi^{-1}(F_{X_t}(x_t^-)); \Phi^{-1}(F_{X_t}(x_t))]. \quad (6.2)$$

The construction of the normal pseudo-residual segment of a discrete
random variable is illustrated in Figure 6.4.

Both versions of pseudo-residual segments (uniform and normal) contain information on how extreme and how rare the observations are,
although the uniform version represents the rarity or otherwise more
directly, as the length of the segment is the corresponding probability.
For example, the lower limit u_t^- of the uniform pseudo-residual interval specifies the probability of observing a value strictly less than x_t,
$1 - u_t^+$ gives the probability of a value strictly greater than x_t, and the
difference $u_t^+ - u_t^-$ is equal to the probability of the observation x_t under the fitted model. The pseudo-residual segments can be interpreted
as interval-censored realizations of a uniform (or standard normal) distribution, if the fitted model is valid. Though this is correct only if the
parameters of the fitted model are known, it is still approximately correct
if the number of estimated parameters is small compared to the size of
the sample (Stadie, 2002). Diagnostic plots of pseudo-residual segments
of discrete random variables necessarily look rather different from those
of continuous random variables.

It is easy to construct an index plot of pseudo-residual segments or to plot these against any independent or dependent variable. However, in order to construct a qq-plot of the pseudo-residual segments one has to specify an ordering of the pseudo-residual segments. One possibility is to sort on the so-called 'mid-pseudo-residuals' which are defined as

$$z_t^m = \Phi^{-1} \left(\frac{u_t^- + u_t^+}{2} \right). \tag{6.3}$$

Furthermore, the mid-pseudo-residuals can themselves be used for checking for normality, for example via a histogram of mid-pseudo-residuals.

Now, having outlined the properties of pseudo-residuals, we can consider the use of pseudo-residuals in the context of HMMs. The analysis of the pseudo-residuals of an HMM serves two purposes: the assessment of the general fit of a selected model, and the detection of outliers. Depending on the aspects of the model that are to be analysed, one can distinguish two kinds of pseudo-residual that are useful for an HMM: those that are based on the conditional distribution given all other observations, which we call **ordinary pseudo-residuals**, and those based on the conditional distribution given all preceding observations, which we call **forecast pseudo-residuals**.

That the pseudo-residuals of a set of observations are identically distributed (either U(0,1) or standard normal) is their crucial property. But for our purposes it is not important whether such pseudo-residuals are independent of each other; indeed we shall see in Section 6.3.2 that it would be wrong to assume of ordinary pseudo-residuals that they are independent.

Note that Dunn and Smyth (1996) discuss (under the name 'quantile residual') what we have called normal pseudo-residuals, and point out that they are a case of Cox–Snell residuals (Cox and Snell, 1968).

6.2.2 Ordinary pseudo-residuals

The first technique considers the observations one at a time and seeks those which, relative to the model and *all* other observations in the series, are sufficiently extreme to suggest that they differ in nature or origin from the others. This means that one computes a pseudo-residual z_t from the conditional distribution of X_t, given $\mathbf{X}^{(-t)}$; a 'full conditional distribution', in the terminology used in MCMC (Markov chain Monte Carlo). For continuous observations the normal pseudo-residual is

$$z_t = \Phi^{-1} \left(\Pr(X_t \leq x_t \mid \mathbf{X}^{(-t)} = \mathbf{x}^{(-t)}) \right).$$

If the model is correct, z_t is a realization of a standard normal random variable. For discrete observations the normal pseudo-residual segment is $[z_t^-; z_t^+]$, where

$$z_t^- = \Phi^{-1}\left(\Pr(X_t < x_t \mid \mathbf{X}^{(-t)} = \mathbf{x}^{(-t)})\right)$$

and

$$z_t^+ = \Phi^{-1}\left(\Pr(X_t \leq x_t \mid \mathbf{X}^{(-t)} = \mathbf{x}^{(-t)})\right).$$

In the discrete case the conditional probabilities $\Pr(X_t = x \mid \mathbf{X}^{(-t)} = \mathbf{x}^{(-t)})$ are given by Equations (5.1) and (5.2) in Section 5.1; the continuous case is similar, with probabilities replaced by densities.

Section 6.3.1 applies ordinary pseudo-residuals to HMMs with one to four states fitted to the earthquakes data, and a further example of their use appears in Figure 9.3 and the corresponding text.

6.2.3 Forecast pseudo-residuals

The second technique for outlier detection seeks observations that are extreme relative to the model and all *preceding* observations (as opposed to all other observations). In this case the relevant conditional distribution is that of X_t given $\mathbf{X}^{(t-1)}$. The corresponding (normal) pseudo-residuals are

$$z_t = \Phi^{-1}\left(\Pr(X_t \leq x_t \mid \mathbf{X}^{(t-1)} = \mathbf{x}^{(t-1)})\right)$$

for continuous observations; and $[z_t^-; z_t^+]$ for discrete, where

$$z_t^- = \Phi^{-1}\left(\Pr(X_t < x_t \mid \mathbf{X}^{(t-1)} = \mathbf{x}^{(t-1)})\right)$$

and

$$z_t^+ = \Phi^{-1}\left(\Pr(X_t \leq x_t \mid \mathbf{X}^{(t-1)} = \mathbf{x}^{(t-1)})\right).$$

In the discrete case the required conditional probability $\Pr(X_t = x_t \mid \mathbf{X}^{(t-1)} = \mathbf{x}^{(t-1)})$ is given by the ratio of the likelihood of the first t observations to that of the first $t-1$:

$$\Pr(X_t = x \mid \mathbf{X}^{(t-1)} = \mathbf{x}^{(t-1)}) = \frac{\boldsymbol{\alpha}_{t-1}\boldsymbol{\Gamma}\mathbf{P}(x)\mathbf{1}'}{\boldsymbol{\alpha}_{t-1}\mathbf{1}'}.$$

The pseudo-residuals of this second type are described as forecast pseudo-residuals because they measure the deviation of an observation from the median of the corresponding one-step-ahead forecast. If a forecast pseudo-residual is extreme, this indicates that the observation concerned is an outlier, or that the model no longer provides an acceptable description of the series. This provides a method for the continuous monitoring of the behaviour of a time series. An example of such monitoring is given at the end of Section 15.4: see Figure 15.5.

The idea of forecast pseudo-residual appears — as 'conditional quantile residual' — in Dunn and Smyth (1996); in the last paragraph on

p. 243 they point out that the quantile residuals they describe can be extended to serially dependent data. The basic idea of (uniform) forecast pseudo-residuals goes back to Rosenblatt (1952), however. Both Brockwell (2007) and Dunn and Smyth describe a way of extending what we call forecast pseudo-residuals to distributions other than continuous. Instead of using a segment of positive length to represent the residual if the observations are not continuous, they choose a point distributed uniformly on that segment. The use of a segment of positive length has the advantage, we believe, of explicitly displaying the discreteness of the observation, and indicating both its extremeness and its rarity.

Another example of the use of forecast pseudo-residuals appears in Figure 9.4 and the corresponding text.

6.3 Examples

6.3.1 Ordinary pseudo-residuals for the earthquakes

In Figure 6.5 we show several types of residual plot for the fitted models of the earthquakes series, using the first definition of pseudo-residual, that based on the conditional distribution relative to all other observations, $\Pr(X_t = x \mid \mathbf{X}^{(-t)} = \mathbf{x}^{(-t)})$. The relevant code appears in A.2.10. It is interesting to compare the pseudo-residuals of the selected three-state model to those of the models with one, two and four states.

As regards the residual plots provided in Figure 6.5, it is clear that the selected three-state model provides an acceptable fit while, for example, the normal pseudo-residuals of the one-state model (a single Poisson distribution) deviate strikingly from the standard normal distribution. If, however, we consider only the residual plots (other than perhaps the qq-plot), and not the model selection criteria, we might even accept the two-state model as an adequate alternative.

Looking at the last row of Figure 6.5, one might be tempted to conjecture that, if a model is 'true', the ordinary pseudo-residuals will be independent, or at least uncorrelated. From the example in the next section we shall see that that would be an incorrect conclusion.

6.3.2 Dependent ordinary pseudo-residuals

Consider the stationary Gaussian AR(1) process $X_t = \phi X_{t-1} + \varepsilon_t$, with the innovations ε_t independent standard normal. It follows that $|\phi| < 1$ and $\text{Var}(X_t) = 1/(1 - \phi^2)$.

Let t lie strictly between 1 and T. The conditional distribution of X_t given $\mathbf{X}^{(-t)}$ is that of X_t given only X_{t-1} and X_{t+1}. This latter conditional distribution can be found by noting that the joint distribution of X_t, X_{t-1} and X_{t+1} (in that order) is normal with mean vector $\mathbf{0}$ and

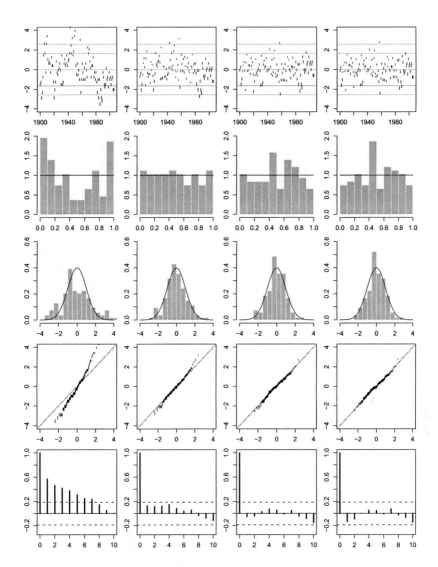

Figure 6.5 *Earthquakes: ordinary pseudo-residuals. Columns 1–4 relate to HMMs with (respectively) 1, 2, 3, 4 states. The top row shows index plots of the normal pseudo-residuals, with horizontal lines at 0, ±1.96, ±2.58. The second and third rows show histograms of the uniform and the normal pseudo-residuals. The fourth row shows quantile-quantile plots of the normal pseudo-residuals, with the theoretical quantiles on the horizontal axis. The last row shows the autocorrelation functions of the normal pseudo-residuals.*

covariance matrix

$$\Sigma = \frac{1}{1 - \phi^2} \begin{pmatrix} 1 & \phi & \phi \\ \phi & 1 & \phi^2 \\ \phi & \phi^2 & 1 \end{pmatrix}.$$

The required conditional distribution then turns out to be normal with mean $\phi(X_{t-1} + X_{t+1})/(1 + \phi^2)$ and variance $(1 + \phi^2)^{-1}$. Hence the corresponding uniform pseudo-residual is

$$\Phi\left(\left(X_t - \frac{\phi}{1 + \phi^2}(X_{t-1} + X_{t+1}) \right) \sqrt{1 + \phi^2} \right),$$

and the normal pseudo-residual is

$$z_t = \left(X_t - \frac{\phi}{1 + \phi^2}(X_{t-1} + X_{t+1}) \right) \sqrt{1 + \phi^2}.$$

By the properties of ordinary normal pseudo-residuals (see Section 6.2.2), z_t is unconditionally standard normal; this can also be verified directly.

Whether (e.g.) the pseudo-residuals z_t and z_{t+1} are independent is not immediately obvious. The answer is that they are not independent; the correlation of z_t and z_{t+1} can (for $t > 1$ and $t < T - 1$) be obtained by routine manipulation, and turns out to be

$$-\phi(1 - \phi^2)/(1 + \phi^2);$$

see Exercise 3. This is opposite in sign to ϕ and smaller in modulus. For instance, if $\phi = 1/\sqrt{2}$, the correlation of z_t and z_{t+1} is $-\phi/3$. (In contrast, the corresponding *forecast* pseudo-residuals do have zero correlation at lag 1.)

One can, also by routine manipulation, show that $\mathrm{Cov}(z_t, z_{t+2}) = 0$ and that, for all integers $k \geq 3$,

$$\mathrm{Cov}(z_t, z_{t+k}) = \phi\,\mathrm{Cov}(z_t, z_{t+k-1}).$$

Consequently, for all integers $k \geq 2$,

$$\mathrm{Cov}(z_t, z_{t+k}) = 0.$$

□

6.4 Discussion

This may be an appropriate point at which to stress the dangers of over-interpretation. Although our earthquakes model seems adequate in important respects, this does not imply that it can be interpreted substantively. Nor indeed are we aware of any convincing seismological interpretation of the three states we propose. But models need not have a substantive interpretation to be useful; many useful statistical models are merely empirical models, in the sense in which Cox (1990) uses that term.

Latent-variable models of all kinds, including independent mixtures and HMMs, seem to be particularly prone to over-interpretation, and we would caution against the error of reification: the tendency to regard as physically real anything that has been given a name. In this spirit we can do no better than to follow Gould (1997, p. 350) in quoting John Stuart Mill:

> The tendency has always been strong to believe that whatever received a name must be an entity or being, having an independent existence of its own. And if no real entity answering to the name could be found, men did not for that reason suppose that none existed, but imagined that it was something peculiarly abstruse and mysterious.

Exercises

1.(a) Let X be a continuous random variable with distribution function F. Show that the random variable $U = F(X)$ is uniformly distributed on the interval $[0, 1]$, i.e. $U \sim U(0, 1)$.

 (b) Suppose that $U \sim U(0, 1)$ and let F be the distribution function of a continuous random variable. Show that the random variable $X = F^{-1}(U)$ has the distribution function F.

 (c) i. Give the explicit expression for F^{-1} for the exponential distribution, i.e. the distribution with density function $f(x) = \lambda e^{-\lambda x}$, $x \geq 0$.

   ii. Verify your result by generating 1000 uniformly distributed random numbers, transforming these by applying F^{-1}, and then examining the histogram of the resulting values.

 (d) Show that for a continuous random variable X with distribution function F, the random variable $Z = \Phi^{-1}(F(X))$ is distributed standard normal.

2. Consider the AR(1) process in the example of Section 6.3.2. That the conditional distribution of X_t given $\mathbf{X}^{(-t)}$ depends only on X_{t-1} and X_{t+1} is fairly obvious because X_t depends on $X_1, \ldots, X_{t-2}$ only through X_{t-1}, and on $X_{t+2}, \ldots, X_T$ only through X_{t+1}.

 Establish this more formally by writing the densities of $\mathbf{X}^{(T)}$ and $\mathbf{X}^{(-t)}$ in terms of conditional densities $p(x_u \mid x_{u-1})$ and noting that, in their ratio, many of the factors cancel.

3. Verify, in the AR(1) example of Section 6.3.2, that for appropriate ranges of t-values:

(a) given X_{t-1} and X_{t+1},

$$X_t \sim N\Big(\phi(X_{t-1} + X_{t+1})/(1 + \phi^2), (1 + \phi^2)^{-1}\Big);$$

(b) $\text{Var}(z_t) = 1$;

(c) $\text{Corr}(z_t, z_{t+1}) = -\phi(1 - \phi^2)/(1 + \phi^2)$;

(d) $\text{Cov}(z_t, z_{t+2}) = 0$; and

(e) for all integers $k \geq 3$, $\text{Cov}(z_t, z_{t+k}) = \phi \, \text{Cov}(z_t, z_{t+k-1})$.

4. Generate a two-state stationary Poisson–HMM $\{X_t\}$ with t.p.m. $\boldsymbol{\Gamma} = \begin{pmatrix} 0.6 & 0.4 \\ 0.4 & 0.6 \end{pmatrix}$, and with state-dependent means $\lambda_1 = 2$ and $\lambda_2 = 5$ for $t = 1$ to 100, and $\lambda_1 = 2$ and $\lambda_2 = 7$ for $t = 101$ to 120. Fit a model to the first 80 observations, and use forecast pseudo-residuals to monitor the next 40 observations for evidence of a change.

5. Consider again the soap sales series introduced in Exercise 5 of Chapter 1.

(a) Use AIC and BIC to decide how many states are needed in a Poisson–HMM for these data.

(b) Compute the pseudo-residuals relative to Poisson–HMMs with 1–4 states, and use plots similar to those in Figure 6.5 to decide how many states are needed.

CHAPTER 7

Bayesian inference for Poisson–hidden Markov models

As alternative to the frequentist approach, one can also consider Bayesian estimation. There are several approaches to Bayesian inference in hidden Markov models: see for instance Chib (1996), Robert and Titterington (1998), Robert, Rydén and Titterington (2000), Scott (2002), Cappé *et al.* (2005) and Frühwirth-Schnatter (2006). Here we follow Scott (2002) and Congdon (2006).

Our purpose is to demonstrate an application of Bayesian inference to Poisson–HMMs. There are obstacles to be overcome, e.g. label switching and the difficulty of estimating m, the number of states, and some of these are model specific.

7.1 Applying the Gibbs sampler to Poisson–HMMs

We consider here a Poisson–HMM $\{X_t\}$ on m states, with underlying Markov chain $\{C_t\}$. We denote the state-dependent means, as usual, by $\boldsymbol{\lambda} = (\lambda_1, \ldots, \lambda_m)$, and the transition probability matrix of the Markov chain by $\boldsymbol{\Gamma}$.

Given a sequence of observations $x_1, x_2, \ldots, x_T$, a fixed m, and prior distributions on the parameters $\boldsymbol{\lambda}$ and $\boldsymbol{\Gamma}$, our objective in this section is to estimate the posterior distribution of these parameters by means of the Gibbs sampler. We shall later (in Section 7.2) drop the assumption that m is known, and also consider the Bayesian estimation thereof.

The prior distributions we assume for the parameters are of the following forms. The r th row $\boldsymbol{\Gamma}_r$ of the t.p.m. $\boldsymbol{\Gamma}$ is assumed to have a Dirichlet distribution with parameter vector $\boldsymbol{\nu}_r$, and the increment $\tau_j = \lambda_j - \lambda_{j-1}$ (with $\lambda_0 \equiv 0$) to have a gamma distribution with shape parameter a_j and rate parameter b_j. Furthermore, the rows of $\boldsymbol{\Gamma}$ and the quantities τ_j are assumed mutually independent in their prior distributions.

Our notation and terminology are as follows. Random variables Y_1, $\ldots$, Y_m are here said to have a Dirichlet distribution with parameter vector $(\nu_1, \ldots, \nu_m)$ if their joint density is proportional to

$$y_1^{\nu_1-1} y_2^{\nu_2-1} \cdots y_m^{\nu_m-1}.$$

More precisely, this expression, with y_m replaced by $1 - \sum_{i=1}^{m-1} y_i$, is (up to proportionality) the joint density of $Y_1, \ldots, Y_{m-1}$ on the unit simplex in dimension $m - 1$, i.e. on the subspace of $\mathbb{R}^{m-1}$ defined by $\sum_{i=1}^{m-1} y_i \leq 1$, $y_i \geq 0$. A random variable X is said to have a gamma distribution with shape parameter a and rate parameter b if its density is (for positive x)

$$f(x) = \frac{b^a}{\Gamma(a)} x^{a-1} e^{-bx}.$$

With this parametrization, X has mean a/b, variance a/b^2 and coefficient of variation (c.v.) $1/\sqrt{a}$.

If we were to observe the Markov chain, updating the transition probabilities $\boldsymbol{\Gamma}$ would be straightforward. Here, however, we have to generate sample paths of the Markov chain in order to update $\boldsymbol{\Gamma}$.

An important part of Scott's model structure, which we copy, is this. Each observed count x_t is considered to be the sum $\sum_j x_{jt}$ of contributions from up to m regimes, the contribution of regime j to x_t being x_{jt}. Note that, if the Markov chain is in state i at a given time, regimes 1 to i are *all* said to be active at that time, and regimes $i + 1$ to m to be inactive. This is an unusual use of the word 'regime', but convenient here.

Instead of parametrizing the model in terms of the m state-dependent means (in our notation, the quantities λ_i), Scott parametrizes it in terms of nonnegative increments $\boldsymbol{\tau} = (\tau_1, \ldots, \tau_m)$, where $\tau_j = \lambda_j - \lambda_{j-1}$ (with $\lambda_0 \equiv 0$). Equivalently,

$$\lambda_i = \sum_{j=1}^{i} \tau_j.$$

This has the effect of placing the λ_js in increasing order, which is useful in order to prevent the technical problem known as label switching. For an account of this problem, see e.g. Frühwirth-Schnatter (2006, Section 3.5.5). The random variable τ_j can be described as the mean contribution of regime j, if active, to the count observed at a given time.

In outline, we proceed as follows.

- Given the observed counts $\mathbf{x}^{(T)}$ and the current values of the parameters $\boldsymbol{\Gamma}$ and $\boldsymbol{\lambda}$, we generate a sample path of the Markov chain.

- We use this sample path to decompose the observed counts into (simulated) regime contributions.

- With the MC sample path available, and the regime contributions, we can now update $\boldsymbol{\Gamma}$ and $\boldsymbol{\tau}$, hence $\boldsymbol{\lambda}$.

The above steps are repeated a large number of times and, after a 'burn-in period', the resulting samples of values of $\boldsymbol{\Gamma}$ and $\boldsymbol{\lambda}$ provide the required estimates of their posterior distributions. In what follows, we use $\boldsymbol{\theta}$ to represent both $\boldsymbol{\Gamma}$ and $\boldsymbol{\lambda}$.

7.1.1 Generating sample paths of the Markov chain

Given the observations $\mathbf{x}^{(T)}$ and the current values of the parameters $\boldsymbol{\theta}$, we wish to simulate a sample path $\mathbf{C}^{(T)}$ of the Markov chain, from its conditional distribution

$$\Pr(\mathbf{C}^{(T)} \mid \mathbf{x}^{(T)}, \boldsymbol{\theta}) = \Pr(C_T \mid \mathbf{x}^{(T)}, \boldsymbol{\theta}) \times \prod_{t=1}^{T-1} \Pr(C_t \mid \mathbf{x}^{(T)}, \mathbf{C}_{t+1}^T, \boldsymbol{\theta}).$$

We shall be drawing values for C_T, C_{T-1}, ..., C_1, in that order, and quantities that we shall need in order to do so are the probabilities

$$\Pr(C_t \mid \mathbf{x}^{(t)}, \boldsymbol{\theta}) = \frac{\Pr(C_t, \mathbf{x}^{(t)} \mid \boldsymbol{\theta})}{\Pr(\mathbf{x}^{(t)} \mid \boldsymbol{\theta})} = \frac{\alpha_t(C_t)}{L_t} \propto \alpha_t(C_t), \quad \text{for } t = 1, \ldots, T.$$
(7.1)

As before (see p. 59), $\boldsymbol{\alpha}_t = (\alpha_t(1), \ldots, \alpha_t(m))$ denotes the vector of forward probabilities

$$\alpha_t(i) = \Pr(\mathbf{x}^{(t)}, C_t = i),$$

which can be computed from the recursion $\boldsymbol{\alpha}_t = \boldsymbol{\alpha}_{t-1}\boldsymbol{\Gamma}\mathbf{P}(x_t)$ ($t = 2, \ldots, T$), with $\boldsymbol{\alpha}_1 = \boldsymbol{\delta}\mathbf{P}(x_1)$; L_t is the likelihood of the first t observations.

We start the simulation by drawing C_T, the state of the Markov chain at the final time T, from $\Pr(C_T \mid \mathbf{x}^{(T)}, \boldsymbol{\theta}) \propto \alpha_T(C_T)$, (i.e. case $t = T$ of Equation (7.1)). We then simulate the states C_t (in the order $t = T - 1$, $T - 2$, ..., 1) by making use of the following proportionality argument, as in Chib (1996):

$$\Pr(C_t \mid \mathbf{x}^{(T)}, \mathbf{C}_{t+1}^T, \boldsymbol{\theta})$$
$$\propto \quad \Pr(C_t \mid \mathbf{x}^{(t)}, \boldsymbol{\theta}) \Pr(\mathbf{x}_{t+1}^T, \mathbf{C}_{t+1}^T \mid \mathbf{x}^{(t)}, C_t, \boldsymbol{\theta})$$
$$\propto \quad \Pr(C_t \mid \mathbf{x}^{(t)}, \boldsymbol{\theta}) \Pr(C_{t+1} \mid C_t, \boldsymbol{\theta}) \Pr(\mathbf{x}_{t+1}^T, \mathbf{C}_{t+2}^T \mid \mathbf{x}^{(t)}, C_t, C_{t+1}, \boldsymbol{\theta})$$
$$\propto \quad \alpha_t(C_t) \Pr(C_{t+1} \mid C_t, \boldsymbol{\theta}).$$
(7.2)

The third factor appearing in the second-last line is independent of C_t, hence the simplification. (See Exercise 4.) The expression (7.2) is easily available, since the second factor in it is simply a one-step transition probability in the Markov chain. We are therefore in a position to simulate sample paths of the Markov chain, given observations $\mathbf{x}^{(T)}$ and parameters $\boldsymbol{\theta}$.

7.1.2 Decomposing the observed counts into regime contributions

Suppose we have a sample path $\mathbf{C}^{(T)}$ of the Markov chain, generated as described in Section 7.1.1, and suppose that $C_t = i$, so that regimes 1 to i are active at time t. Our next step is to decompose each observation x_t ($t = 1, 2, \ldots, T$) into regime contributions $x_{1t}, \ldots, x_{it}$ such that $\sum_{j=1}^{i} x_{jt} = x_t$. We therefore need the joint distribution of $X_{1t}, \ldots, X_{it}$, given $C_t = i$ and $X_t = x_t$ (and given $\boldsymbol{\theta}$). This is multinomial with total x_t and probability vector proportional to $(\tau_1, \ldots, \tau_i)$; see Exercise 1.

7.1.3 Updating the parameters

The transition probability matrix $\boldsymbol{\Gamma}$ can now be updated, i.e. new estimates produced. This we do by drawing $\boldsymbol{\Gamma}_r$, the rth row of the t.p.m. $\boldsymbol{\Gamma}$, from the Dirichlet distribution with parameter vector $\boldsymbol{\nu}_r + \mathbf{T}_r$, where $\mathbf{T}_r$ is the rth row of the (simulated) matrix of transition counts; see Section 7.1.1. (Recall that the prior for $\boldsymbol{\Gamma}_r$ is Dirichlet($\boldsymbol{\nu}_r$), and see Exercise 2.)

Similarly, the vector $\boldsymbol{\lambda}$ of state-dependent means is updated by drawing τ_j ($j = 1, \ldots, m$) from a gamma distribution with parameters $a_j + \sum_{t=1}^{T} x_{jt}$ and $b_j + N_j$; here N_j denotes the number of times regime j was active in the simulated sample path of the Markov chain, and x_{jt} the contribution of regime j to x_t. (Recall that the prior for τ_j is a gamma distribution with shape parameter a_j and rate parameter b_j, and see Exercise 3.)

7.2 Bayesian estimation of the number of states

In the Bayesian approach to model selection, the number of states, m, is a parameter whose value is assessed from its posterior distribution, $p(m \mid \mathbf{x}^{(T)})$. Computing this posterior distribution is, however, not an easy problem; indeed it has been described as 'notoriously difficult to calculate' (Scott, James and Sugar, 2005).

Using p as a general symbol for probability mass or density functions, one has

$$p(m \mid \mathbf{x}^{(T)}) = p(m)\, p(\mathbf{x}^{(T)} \mid m)/p(\mathbf{x}^{(T)}) \propto p(m)\, p(\mathbf{x}^{(T)} \mid m), \qquad (7.3)$$

where $p(\mathbf{x}^{(T)} \mid m)$ is called the integrated likelihood. If only two models are being compared, the posterior odds are equal to the product of the 'Bayes factor' and the prior odds:

$$\frac{p(m_2 \mid \mathbf{x}^{(T)})}{p(m_1 \mid \mathbf{x}^{(T)})} = \frac{p(\mathbf{x}^{(T)} \mid m_2)}{p(\mathbf{x}^{(T)} \mid m_1)} \times \frac{p(m_2)}{p(m_1)}. \qquad (7.4)$$

7.2.1 Use of the integrated likelihood

In order to use (7.3) or (7.4) we need to estimate the integrated likelihood

$$p(\mathbf{x}^{(T)} \mid m) = \int p(\boldsymbol{\theta}_m, \mathbf{x}^{(T)} \mid m) \, \mathrm{d}\boldsymbol{\theta}_m = \int p(\mathbf{x}^{(T)} \mid m, \boldsymbol{\theta}_m) \, p(\boldsymbol{\theta}_m \mid m) \, \mathrm{d}\boldsymbol{\theta}_m.$$

One way of doing so would be to simulate from $p(\boldsymbol{\theta}_m \mid m)$, the prior distribution of the parameters $\boldsymbol{\theta}_m$ of the m-state model. But it is convenient and — especially if the prior is diffuse — more efficient to use a method that requires instead a sample from the posterior distribution, $p(\boldsymbol{\theta}_m \mid \mathbf{x}^{(T)}, m)$. Such a method is as follows.

Write the integrated likelihood as

$$\int p(\mathbf{x}^{(T)} \mid m, \boldsymbol{\theta}_m) \, \frac{p(\boldsymbol{\theta}_m \mid m)}{p^*(\boldsymbol{\theta}_m)} \, p^*(\boldsymbol{\theta}_m) \, \mathrm{d}\boldsymbol{\theta}_m;$$

i.e. write it in a form suitable for the use of a sample from some convenient density $p^*(\boldsymbol{\theta}_m)$ for the parameters $\boldsymbol{\theta}_m$. Since we have available a sample $\boldsymbol{\theta}_k^{(j)}$ $(j = 1, 2, \ldots, B)$ from the posterior distribution, we can use that sample; i.e. we can take $p^*(\boldsymbol{\theta}_m) = p(\boldsymbol{\theta}_m \mid \mathbf{x}^{(T)}, m)$. Newton and Raftery (1994) therefore suggest *inter alia* that the integrated likelihood can be estimated by

$$\hat{I} = \sum_{j=1}^{B} w_j p(\mathbf{x}^{(T)} \mid m, \boldsymbol{\theta}_m^{(j)}) \bigg/ \sum_{j=1}^{B} w_j, \tag{7.5}$$

where

$$w_j = \frac{p(\boldsymbol{\theta}_m^{(j)} \mid m)}{p(\boldsymbol{\theta}_m^{(j)} \mid \mathbf{x}^{(T)}, m)}.$$

After some manipulation this simplifies to the harmonic mean of the likelihood values of a sample from the posterior:

$$\hat{I} = \left(B^{-1} \sum_{j=1}^{B} p(\mathbf{x}^{(T)} \mid m, \boldsymbol{\theta}_m^{(j)}) \right)^{-1}; \tag{7.6}$$

see Exercise 5 for the details. This is, however, not the only route one can follow in deriving the estimator (7.6); see Exercise 6 for another possibility.

Newton and Raftery state that, under quite general conditions, $\hat{I}$ is a simulation-consistent estimator of $p(\mathbf{x}^{(T)} \mid m)$. But there is a major drawback to this harmonic mean estimator, its infinite variance, and the question of which estimator to use for $p(\mathbf{x}^{(T)} \mid m)$ does not seem to have been settled. Raftery *et al.* (2007) suggest two alternatives to the harmonic mean estimator, but no clear recommendation emerges, and one of the discussants of that paper (Draper, 2007) bemoans the disheartening 'ad-hockery' of the many proposals that have over the years been made for coping with the instability of expectations with respect to (often diffuse) priors.

7.2.2 Model selection by parallel sampling

However, it is possible to estimate $p(m \mid \mathbf{x}^{(T)})$ relatively simply by 'parallel sampling' of the competing models, provided that the set of competing models is sufficiently small; see Congdon (2006) and Scott (2002). Denote by $\boldsymbol{\theta}$ the vector $(\boldsymbol{\theta}_1, \boldsymbol{\theta}_2, \ldots, \boldsymbol{\theta}_K)$, and similarly $\boldsymbol{\theta}^{(j)}$; K is the maximum number of states. Make the assumption that

$$p(m, \boldsymbol{\theta}) = p(\boldsymbol{\theta}_m \mid m)\, p(m);$$

that is, assume that model m is (for $j \neq m$) indifferent to values taken by $\boldsymbol{\theta}_j$.

We wish to estimate $p(m \mid \mathbf{x}^{(T)})$ (for $m = 1, \ldots, K$) by

$$B^{-1} \sum_{j=1}^{B} p(m \mid \mathbf{x}^{(T)}, \boldsymbol{\theta}^{(j)}). \tag{7.7}$$

We use the fact that, with the above assumption,

$$p(m \mid \mathbf{x}^{(T)}, \boldsymbol{\theta}^{(j)}) \propto G_m^{(j)},$$

where

$$G_m^{(j)} \equiv p(\mathbf{x}^{(T)} \mid \boldsymbol{\theta}_m^{(j)}, m)\, p(\boldsymbol{\theta}_m^{(j)} \mid m)\, p(m). \tag{7.8}$$

(See Appendix 1 of Congdon (2006).) Hence

$$p(m \mid \mathbf{x}^{(T)}, \boldsymbol{\theta}^{(j)}) = G_m^{(j)} \Big/ \sum_{k=1}^{K} G_k^{(j)}.$$

This expression for $p(m \mid \mathbf{x}^{(T)}, \boldsymbol{\theta}^{(j)})$ can then be inserted in (7.7) to complete the estimate of $p(m \mid \mathbf{x}^{(T)})$.

7.3 Example: earthquakes

We apply the techniques described above to the series of annual counts of major earthquakes. The prior distributions used are as follows. The gamma distributions used as priors for the λ-increments (i.e. for the quantities τ_j) all have mean $50m/(m+1)$ and c.v. 1 in one analysis, and 2 in a second. The Dirichlet distributions used as priors for the rows of $\boldsymbol{\Gamma}$ all have all parameters equal to 1. The prior distribution for m, the number of states, assigns probability $\frac{1}{6}$ to each of the values $1, 2, \ldots, 6$. The number of iterations used was $B = 100\,000$, with a burn-in period of 5000.

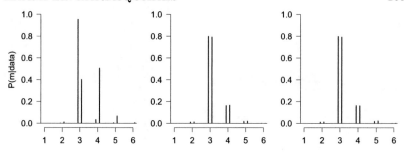

Figure 7.1 *Earthquakes data, posterior distributions of m given a uniform prior on* $\{1, 2, \ldots, 6\}$. *Each panel shows two posterior distributions from independent runs. In the left and centre panels the c.v. in the gamma prior is 1. Left panel: harmonic mean estimator. Centre panel: parallel sampling estimator. Right panel: parallel sampling estimator with c.v. = 1 (left bars) and c.v. = 2 (right bars).*

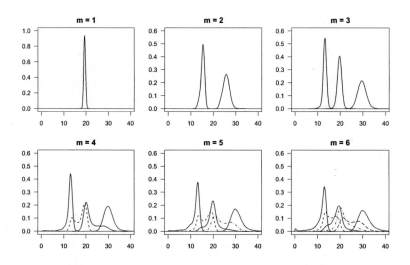

Figure 7.2 *Earthquakes data: posterior distributions of the state-dependent means for one- to six-state Poisson–HMMs.*

Figure 7.1 displays three comparisons of estimates of the posterior distribution of m. It is clear from the comparison of two independent runs of the harmonic mean estimator (left panel) that it is indeed very unstable. In the first run it would have chosen a three-state model by a large margin, and in the second run a four-state model. In contrast, the parallel sampling estimator (centre and right panels) produces very

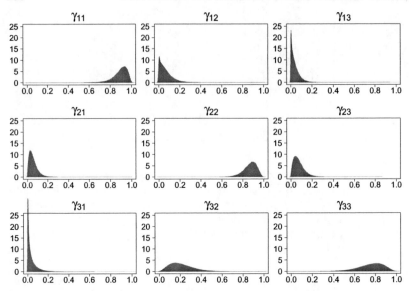

Figure 7.3 *Three-state Poisson–HMM for earthquakes data: posterior distribution of transition probability matrix* Γ.

consistent results even if the c.v. of the prior distributions for the λ-increments is changed from 1 to 2, and clearly identifies $m = 3$ as the posterior mode.

The posterior distributions of the Poisson means for $m = 1, \ldots, 6$ are displayed in Figure 7.2. The posterior distributions of the entries of Γ for the three-state model are displayed in Figure 7.3. Table 7.1 lists posterior statistics for the three-state model. The posterior modes are generally quite close to the maximum likelihood estimates given on p. 51. In particular the values λ are almost the same, but the posterior modes for the entries of Γ are mostly a little closer to 0.5 than are the corresponding MLEs.

7.4 Discussion

From the above example it seems clear that the Bayesian approach is demanding computationally and in certain other respects. The model needs to be parametrized in a way that avoids label switching. In HMMs the labels associated with the states are arbitrary; the model is invariant under permutation of the labels. This is irrelevant to maximum likelihood estimation, but it is a problem in the context of MCMC estimation of posterior distributions. One must ensure that only one of the $m!$ per-

Table 7.1 *Earthquakes data: posterior statistics for the three-state model.*

parameter	min	Q1	mode	median	mean	Q3	max
λ_1	6.21	12.62	13.12	13.15	13.12	13.68	16.85
λ_2	13.53	19.05	19.79	19.74	19.71	20.42	27.12
λ_3	22.08	28.33	29.87	29.59	29.64	30.88	43.88
γ_{11}	0.001	0.803	0.882	0.861	0.843	0.905	0.998
γ_{12}	0.000	0.047	0.056	0.085	0.104	0.139	0.964
γ_{13}	0.000	0.020	0.011	0.042	0.053	0.075	0.848
γ_{21}	0.000	0.043	0.050	0.070	0.083	0.108	0.979
γ_{22}	0.009	0.784	0.858	0.837	0.824	0.880	0.992
γ_{23}	0.000	0.052	0.060	0.082	0.093	0.122	0.943
γ_{31}	0.000	0.021	0.011	0.049	0.068	0.096	0.758
γ_{32}	0.000	0.144	0.180	0.213	0.229	0.296	0.918
γ_{33}	0.010	0.627	0.757	0.718	0.703	0.795	0.986

mutations of the labels is used in the simulation. In the algorithm for
Poisson–HMMs discussed above, label switching was avoided by ordering
the states according to the means of the state-dependent distributions.
An analogous reparametrization has been used for the normal case by
Robert and Titterington (1998). Prior distributions need to be specified for the parameters and, as a rule, the choice of prior distribution
is driven by mathematical convenience rather than by prior information. The above difficulties are model specific; the derivations, priors and
hence the computer code all change substantially if the state-dependent
distribution changes.

Although the computational demands of MCMC are high in comparison with those of ML *point estimation*, this is not a fair comparison. Interval estimation using the parametric bootstrap, though easy
to implement (see Section 3.6.2), is comparably time-consuming. Interval estimates based instead on approximate standard errors obtained
from the Hessian (as in Section 3.6.1) require potentially demanding
parametrization-specific derivations.

A warning note regarding MCMC has been sounded by Celeux, Hurn
and Robert (2000), and echoed by Chopin (2007):

> [. . .] we consider that almost the entirety of MCMC samplers implemented
> for mixture models has failed to converge!

This statement was made of independent mixture models but is presum-
ably also applicable to HMMs.

Another Bayesian approach to model selection in HMMs is the use of reversible jump Markov chain Monte Carlo techniques (RJMCMC), in which m, the number of states, is treated as a parameter of the model and, like the other parameters, updated in each iteration. This has the advantage over parallel sampling that relatively few iterations are 'wasted' on unpromising values of m, thereby reducing the total number of iterations needed to achieve a given degree of accuracy. Such an advantage might be telling for very long time series and a potentially large number of states, but for typical applications and sample sizes, models with $m > 5$ are rarely feasible. The disadvantage for users who have to write their own code is the complexity of the algorithm.

Robert, Rydén and Titterington (2000) describe in detail the use of that approach in HMMs with normal distributions as the state-dependent distributions, and provide several examples of its application. As far as we are aware, this approach has not been extended to HMMs with other possible state-dependent distributions, e.g. Poisson, and we refer the reader interested in RJMCMC to Robert *et al.* and the references therein, especially Green (1995) and Richardson and Green (1997).

Exercises

1. Consider u defined by $u = \sum_{j=1}^{i} u_j$, where the variables u_j are independent Poisson random variables with means τ_j.

   Show that, conditional on u, the joint distribution of $u_1, u_2, \ldots, u_i$ is multinomial with total u and probability vector $(\tau_1, \ldots, \tau_i) / \sum_{j=1}^{i} \tau_j$.

2. (Updating of Dirichlet distributions) Let $\mathbf{w} = (w_1, w_2, \ldots, w_m)$ be an observation from a multinomial distribution with probability vector $\mathbf{y}$, which has a Dirichlet distribution with parameter vector $\mathbf{d} = (d_1, d_2, \ldots, d_m)$.

   Show that the posterior distribution of $\mathbf{y}$, i.e. the distribution of $\mathbf{y}$ given $\mathbf{w}$, is the Dirichlet distribution with parameters $\mathbf{d} + \mathbf{w}$.

3. (Updating of gamma distributions) Let $y_1, y_2, \ldots, y_n$ be a random sample from the Poisson distribution with mean τ, which is gamma-distributed with parameters a and b.

   Show that the posterior distribution of τ, i.e. the distribution of τ given $y_1, \ldots, y_n$, is the gamma distribution with parameters $a + \sum_{i=1}^{n} y_i$ and $b + n$.

4. Show that, in the basic HMM,

$$\Pr(\mathbf{X}_{t+1}^T, \mathbf{C}_{t+2}^T \mid \mathbf{X}^{(t)}, C_t, C_{t+1}) = \Pr(\mathbf{X}_{t+1}^T, \mathbf{C}_{t+2}^T \mid C_{t+1}).$$

   (Hint: either use the methods of Appendix B or invoke d-separation, for which see e.g. Pearl (2000), pp. 16–18.)

5. Consider the estimator $\hat{I}$ as defined by Equation (7.5):

$$\hat{I} = \sum_j w_j \, p(\mathbf{x}^{(T)} \mid m, \boldsymbol{\theta}_m^{(j)}) \bigg/ \sum_j w_j.$$

(a) Show that the weight $w_j = p(\boldsymbol{\theta}_m^{(j)} \mid m)/p(\boldsymbol{\theta}_m^{(j)} \mid \mathbf{x}^{(T)}, m)$ can be written as $p(\mathbf{x}^{(T)} \mid m)/p(\mathbf{x}^{(T)} \mid m, \boldsymbol{\theta}_m^{(j)})$.

(b) Deduce that the summand $w_j \, p(\mathbf{x}^{(T)} \mid m, \boldsymbol{\theta}_m^{(j)})$ is equal to $p(\mathbf{x}^{(T)} \mid m)$ (and is therefore independent of j).

(c) Hence show that $\hat{I}$ is the harmonic mean displayed in Equation (7.6).

6. Let the observations $\mathbf{x}$ be distributed with parameter (vector) $\boldsymbol{\theta}$. Prove that

$$\frac{1}{p(\mathbf{x})} = \mathrm{E}\left(\frac{1}{p(\mathbf{x} \mid \boldsymbol{\theta})} \,\bigg|\, \mathbf{x}\right);$$

i.e. the integrated likelihood $p(\mathbf{x})$ is the harmonic mean of the likelihood $p(\mathbf{x} \mid \boldsymbol{\theta})$ computed under the posterior distribution $p(\boldsymbol{\theta} \mid \mathbf{x})$ for $\boldsymbol{\theta}$. (This is the 'harmonic mean identity', as in Equation (1) of Raftery et al. (2007), and suggests the use of the harmonic mean of likelihoods sampled from the posterior as estimator of the integrated likelihood.)

CHAPTER 8

Extensions of the basic hidden Markov model

A second principle (which applies also to artists!) is not to fall in love with one model to the exclusion of alternatives.

McCullagh and Nelder
Generalized Linear Models (1989, p. 8)

8.1 Introduction

A notable advantage of HMMs is the ease with which the basic model can be modified or generalized, in several different directions, in order to provide flexible models for a wide range of types of observation.

We begin this chapter by describing the use in the basic HMM of univariate state-dependent distributions other than the Poisson (Section 8.2), and then show how the basic HMM can be extended in other ways. The first such extension (Section 8.3) adds flexibility by generalizing the underlying parameter process; the assumption that the parameter process is a first-order Markov chain is relaxed by allowing it to be a second-order Markov chain. This extension can be applied not only to the basic model but also to most of the other models to be discussed.

We then illustrate how the basic model can be generalized to construct HMMs for a number of different and more complex types of observation, including the following.

- Series of multinomial-like observations (Section 8.4.1): An example of a multinomial-like series would be daily sales of a particular item categorized into the four consumer categories: adult female, adult male, juvenile female, juvenile male.

- Categorical series (Section 8.4.2): An important special case of the multinomial-like series is that in which there is exactly one observation at each time, classified into one of q possible mutually exclusive categories: that is, a categorical time series. An example is an hourly series of wind directions in the form of the conventional 16 categories, i.e. the 16 points of the compass.

- Other multivariate series (Section 8.4.3): An example of a bivariate discrete-valued series is the number of sales of each of two related items. A key feature of multivariate time series is that, in addition to serial dependence within each series, there may be dependence across the series.

- Series that depend on covariates (Section 8.5): Many, if not most, time series studied in practice exhibit time trend, seasonal variation, or both. Examples include monthly sales of items, daily number of shares traded, insurance claims received, and so on. One can regard such series as depending on time as a covariate. In some cases covariates other than time are relevant. For example, one might wish to model the number of sales of an item as a function of the price, advertising expenditure and sales promotions, and also to allow for trend and seasonal fluctuations.

- Models with additional dependencies (Section 8.6): One way to produce useful generalizations of the basic HMM as depicted in Figure 2.2 is to add extra dependencies between some of the random variables that make up the model. There are several different ways to do this. For instance, if one suspects that the continuous observations X_{t-1} and X_t are not conditionally independent given the Markov chain, one might wish to use a model that switches between AR(1) processes according to a Markov chain, that is, a Markov-switching AR(1).

8.2 HMMs with general univariate state-dependent distribution

In this book we have introduced the basic (univariate) HMM by concentrating on Poisson state-dependent distributions. One may, however, use any distribution — discrete, continuous, or a mixture of the two — as the state-dependent distribution; in fact there is nothing preventing one from using a different family of distributions for each state. One simply redefines the diagonal matrices containing the state-dependent probabilities, and in the estimation process takes note of whatever constraints the state-dependent parameters must observe, either by transforming the constraints away or by explicitly constrained optimization.

In what follows we describe HMMs with various univariate state-dependent distributions without going into much detail.

- HMMs for unbounded counts

  The Poisson distribution is the canonical model for unbounded counts. However, a popular alternative, especially for overdispersed data, is the negative binomial distribution. One can therefore consider replacing the Poisson state-dependent distribution in a Poisson–HMM by the negative binomial, which is given, for all nonnegative integers x, by the probability function

$$p_i(x) = \frac{\Gamma\left(x + \frac{1}{\eta_i}\right)}{\Gamma\left(\frac{1}{\eta_i}\right)\Gamma(x+1)} \left(\frac{1}{1 + \eta_i \mu_i}\right)^{\frac{1}{\eta_i}} \left(\frac{\eta_i \mu_i}{1 + \eta_i \mu_i}\right)^x,$$

where the parameters μ_i (the mean) and η_i are positive. (But note that this is only one of several possible parametrizations of the negative binomial.) A negative binomial–HMM may sensibly be used if even a Poisson–HMM seems unable to accommodate the observed overdispersion. Conceivable examples for the application of Poisson– or negative binomial–HMMs include series of counts of stoppages or breakdowns of technical equipment, earthquakes, sales, insurance claims, accidents reported, defective items and stock trades.

- HMMs for binary data

  The Bernoulli–HMM for binary time series is the simplest HMM. Its state-dependent probabilities for the two possible outcomes are, for some probabilities π_i, just

$$p_i(0) = \Pr(X_t = 0 \mid C_t = i) = 1 - \pi_i \qquad \text{(failure)},$$
$$p_i(1) = \Pr(X_t = 1 \mid C_t = i) = \pi_i \qquad \text{(success)}.$$

An example of a Bernoulli–HMM appears in Section 2.3.1. Possible applications of Bernoulli–HMMs are to daily rainfall occurrence (rain or no rain), daily trading of a share (traded or not traded), and consecutive departures of aeroplanes from an airport (on time, not on time).

- HMMs for bounded counts

  Binomial–HMMs may be used to model series of bounded counts. The state-dependent binomial probabilities are given by

$$_t p_i(x_t) = \binom{n_t}{x_t} \pi_i^{x_t} (1 - \pi_i)^{n_t - x_t},$$

where n_t is the number of trials at time t and x_t the number of successes. (We use the prefix t as far as possible to indicate time-dependence.)

Possible examples for series of bounded counts that may be described by a binomial–HMM are series of:

- purchasing preferences, e.g. $n_t = $ number of purchases of all brands on day t, $x_t = $ number of purchases of brand A on day t;
- sales of newspapers or magazines, e.g. $n_t = $ number available on day t, $x_t = $ number purchased on day t.

Notice, however, that there is a complication when one computes the forecast distribution of a binomial–HMM. Either n_{T+h}, the number of trials at time $T + h$, must be known, or one has to fit a separate model to forecast n_{T+h}. Alternatively, by setting $n_{T+h} = 1$ one can simply compute the forecast distribution of the 'success proportion'.

- HMMs for continuous-valued series

So far, we have primarily considered HMMs with discrete-valued state-dependent component distributions. However, we have also mentioned that it is possible to use continuous-valued component distributions. One simply has to replace the probability functions by the corresponding state-dependent probability density functions.

Important state-dependent distributions for continuous-valued time series are the exponential, Gamma and normal distributions.

Normal–HMMs are sometimes used for modelling share returns series because the observed kurtosis of most such series is greater than 3, the kurtosis of a normal distribution. See Section 13.2 for a multivariate model for returns on four shares. Note that the (continuous) likelihood of a normal–HMM is unbounded; it is possible to increase the likelihood without bound by fixing a state-dependent mean μ_i at one of the observations and letting the corresponding variance σ_i approach zero. In practice, this may or may not lead to problems in parameter estimation. If it does, using the discrete likelihood is advisable; see Section 1.2.3.

8.3 HMMs based on a second-order Markov chain

One generalization of the basic HMM is that which results if one replaces the underlying first-order Markov chain in the basic model (or in the extensions to follow later in this chapter) by a higher-order chain. Here we describe only the model that has as parameter process a stationary second-order chain. Such a second-order chain is characterized by the transition probabilities

$$\gamma(i, j, k) = P(C_t = k \mid C_{t-1} = j, C_{t-2} = i),$$

and has stationary bivariate distribution $u(j, k) = P(C_{t-1} = j, C_t = k)$. (See Section 1.3.6 for a more detailed description of higher-order Markov chains.)

We mention here two important aspects of such a second-order HMM, which is depicted in Figure 8.1. The first is that it is possible to evaluate the likelihood of a second-order HMM in very similar fashion to that of the basic model; the computational effort is in this case cubic in m, the number of states, and, as before, linear in T, the number of observations.

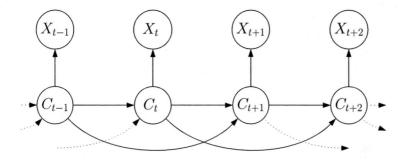

Figure 8.1 *Directed graph of second-order HMM.*

(See Exercise 3.) This then enables one to estimate the parameters by direct maximization of the likelihood, and also to compute the forecast distributions.

The second aspect is the number of free parameters of the (second-order) Markov chain component of the model. In general this number is $m^2(m-1)$, which rapidly becomes prohibitively large as m increases. This overparametrization can be circumvented by using some restricted subclass of second-order Markov chain models, for example those of Pegram (1980) or those of Raftery (1985a). Such models are necessarily less flexible than the general class of second-order Markov chains. They maintain the second-order structure of the chain, but trade some flexibility in return for a reduction in the number of parameters.

Having pointed out that it is possible to increase the order of the Markov chain in an HMM from one to two (or higher), in what follows we shall restrict our attention almost entirely to the simpler case of a first-order Markov chain. The exception is Section 10.2.2, where we present (among other models) an example of a second-order HMM for a binary time series. We note in passing that the complications caused by the constraints on the parameters in a Raftery model have been discussed by Schimert (1992) and by Raftery and Tavaré (1994), who describe how one can reduce the number of constraints.

8.4 HMMs for multivariate series

8.4.1 Series of multinomial-like observations

One of the variations of the basic model mentioned in Section 8.2 is the binomial–HMM, in which, conditional on the Markov chain, the observations $\{x_t : t = 1, \ldots, T\}$ are the numbers of successes in $n_1, n_2, \ldots,$

n_T independent Bernoulli trials. The m-state model has m probabilities of success π_i, one for each state i.

A **multinomial–HMM** is the obvious generalization thereof to the situation in which there are $q \geq 2$, rather than two, mutually exclusive and exhaustive possible outcomes to each trial. The observations are then q series of counts, $\{x_{tj} : t = 1, \ldots, T, \ j = 1, \ldots, q\}$ with $x_{t1} + x_{t2} + \ldots + x_{tq} = n_t$ where n_t is the (known) number of trials at time t. Thus for example x_{23} represents the number of outcomes at time $t = 2$ that were of type 3. The counts x_{tj} at time t can be combined in a vector $\mathbf{x}_t = (x_{t1}, x_{t2}, \ldots, x_{tq})$.

We shall suppose that, conditional on the Markov chain $\mathbf{C}^{(T)}$, the T random vectors $\{\mathbf{X}_t = (X_{t1}, X_{t2}, \ldots, X_{tq}) : t = 1, \ldots, T\}$ are mutually independent.

The parameters of the model are as follows. As in the basic model, the matrix $\boldsymbol{\Gamma}$ has $m(m-1)$ free parameters. With each of the m states of the Markov chain there is associated a multinomial distribution with parameters n_t (known) and q unknown probabilities which, for state i, we shall denote by $\pi_{i1}, \pi_{i2}, \ldots, \pi_{iq}$. These probabilities are constrained by $\sum_{j=1}^{q} \pi_{ij} = 1$ for each state i. This component of the model therefore has $m(q-1)$ free parameters, and the entire model has $m^2 - m + m(q-1) = m^2 + m(q-2)$.

The likelihood of observations $\mathbf{x}_1, \ldots, \mathbf{x}_T$ from a general multinomial–HMM differs little from that of a binomial–HMM; the only difference is that the binomial probabilities

$$_t p_i(x_t) = \binom{n_t}{x_t} \pi_i^{x_t} (1 - \pi_i)^{n_t - x_t}$$

are replaced by multinomial probabilities

$$_t p_i(\mathbf{x}_t) = \mathrm{P}(\mathbf{X}_t = \mathbf{x}_t \mid C_t = i) = \binom{n_t}{x_{t1}, x_{t2}, \ldots, x_{tq}} \pi_{i1}^{x_{t1}} \pi_{i2}^{x_{t2}} \cdots \pi_{iq}^{x_{tq}}.$$

Note that these probabilities are indexed by the time t because the number of trials n_t is permitted to be time-dependent. We assume that the mq state-dependent probabilities π_{ij} are constant over time, but that is an assumption that can if necessary be relaxed.

The likelihood is given by

$$L_T = \boldsymbol{\delta} \,_1\mathbf{P}(\mathbf{x}_1) \boldsymbol{\Gamma} \,_2\mathbf{P}(\mathbf{x}_2) \cdots \boldsymbol{\Gamma} \,_T\mathbf{P}(\mathbf{x}_T) \mathbf{1}',$$

where $_t\mathbf{P}(\mathbf{x}_t) = \mathrm{diag}\,(_t p_1(\mathbf{x}_t), \ldots, {}_t p_m(\mathbf{x}_t))$. Parameters can then be estimated by maximizing the likelihood as a function of $m(q-1)$ of the 'success probabilities', e.g. π_{ij} for $j = 1, 2, \ldots, q-1$, and of the $m^2 - m$ off-diagonal transition probabilities. If one does so, one must observe not only the usual 'generalized upper bound' constraints $\sum_{j \neq i} \gamma_{ij} \leq 1$ on the transition probabilities, but also the m similar constraints $\sum_{j=1}^{q-1} \pi_{ij} \leq 1$ on the probabilities π_{ij}, one constraint for each state i — as well as, of course, the lower bound of 0 on all these probabilities.

Once the parameters have been estimated, these can be used to estimate various forecast distributions. There is, however, the same complication to such forecasts as described already in the case of the binomial–HMM. We have assumed that n_t, the number of trials at time t, is known. This number, being the sum of the q observed counts at time t, is certainly known at times $t = 1, 2, \ldots, T$. But in order to compute the one-step-ahead forecast distribution, one needs to know n_{T+1}, the number of trials that will take place at time $T + 1$. This will be known in some applications, for instance when the number of trials is prescribed by a sampling scheme. But there are also applications in which n_{T+1} is a random variable whose value remains unknown until time $T + 1$. For the latter it is not possible to compute the forecast distribution of the counts at time $T + 1$. As before, by setting $n_{T+1} = 1$ it is possible to compute the forecast distribution of the count-proportions.

An alternative approach is to fit a separate model to the series $\{n_t\}$, to use that model to compute the forecast distribution of n_{T+1} and then, finally, to use that to compute the required forecast distribution for the counts of the multinomial–HMM.

8.4.2 A model for categorical series

A simple but important special case of the multinomial–HMM is that in which $n_t = 1$ for all t. This provides a model for categorical series, e.g. DNA base sequences or amino-acid sequences, in which there is exactly one symbol at each position in the sequence: one of A, C, G, T in the former example, one of 20 amino-acids in the latter. In this case the state-dependent probabilities $_t p_i(\mathbf{x})$ and the matrix expression for the likelihood simplify somewhat.

Because n_t is constant, the prefix t is no longer necessary, and because $\sum_{k=1}^{q} x_{tk} = 1$, the q-vector $\mathbf{x}_t$ has one entry equal to 1 and the others equal to zero. It follows that, if

$$\mathbf{x} = (\underbrace{0, \ldots, 0}_{j-1}, 1, \underbrace{0, \ldots, 0}_{q-j}),$$

then $p_i(\mathbf{x}) = \pi_{ij}$ and

$$\mathbf{P}(\mathbf{x}) = \mathrm{diag}\left(\pi_{1j}, \ldots, \pi_{mj}\right).$$

For convenience we denote $\mathbf{P}(\mathbf{x})$, for $\mathbf{x} = (0, \ldots, 0, 1, 0, \ldots, 0)$ as above, by $\mathbf{\Pi}(j)$. In this notation the likelihood of observing categories $j_1, j_2,$

$\ldots, j_T$ at times $1, 2, \ldots, T$ is given by

$$L_T = \boldsymbol{\delta}\boldsymbol{\Pi}(j_1)\boldsymbol{\Gamma}\boldsymbol{\Pi}(j_2)\boldsymbol{\Gamma}\cdots\boldsymbol{\Pi}(j_T)\mathbf{1}'.$$

If we assume the Markov chain is stationary, it is implied, for instance, that the probability of observing category l at time $t + 1$, given that category k is observed at time t, is

$$\frac{\boldsymbol{\delta}\boldsymbol{\Pi}(k)\boldsymbol{\Gamma}\boldsymbol{\Pi}(l)\mathbf{1}'}{\boldsymbol{\delta}\boldsymbol{\Pi}(k)\mathbf{1}'}, \tag{8.1}$$

and similarly, that of observing l at time $t + 1$, given k at time t and j at time $t - 1$, is

$$\frac{\boldsymbol{\delta}\boldsymbol{\Pi}(j)\boldsymbol{\Gamma}\boldsymbol{\Pi}(k)\boldsymbol{\Gamma}\boldsymbol{\Pi}(l)\mathbf{1}'}{\boldsymbol{\delta}\boldsymbol{\Pi}(j)\boldsymbol{\Gamma}\boldsymbol{\Pi}(k)\mathbf{1}'}.$$

The above two expressions can be used to compute forecast distributions. An example of a forecast using Equation (8.1) is given in Section 12.2.2 (see p. 172).

8.4.3 Other multivariate models

The series of multinomial-like counts discussed in the last section are, of course, examples of multivariate series, but with a specific structure. In this section we illustrate how it is possible to develop HMMs for different and more complex types of multivariate series.

Consider q time series $\{(X_{t1}, X_{t2}, \ldots, X_{tq}) : t = 1, \ldots, T\}$ which we shall represent compactly as $\{\mathbf{X}_t, : t = 1, \ldots, T\}$. As we did for the basic HMM, we shall assume that, conditional on $\mathbf{C}^{(T)} = \{C_t : t = 1, \ldots, T\}$, the above random vectors are mutually independent. We shall refer to this property as **longitudinal conditional independence** in order to distinguish it from a different conditional independence that we shall describe later.

To specify an HMM for such a series it is necessary to postulate a model for the distribution of the random vector $\mathbf{X}_t$ in each of the m states of the parameter process. That is, one requires the following probabilities to be specified for $t = 1, 2, \ldots, T$, $i = 1, 2, \ldots, m$, and all relevant x:

$$_t p_i(\mathbf{x}) = \Pr(\mathbf{X}_t = \mathbf{x} \mid C_t = i).$$

(For generality, we keep the time index t here, i.e. we allow the state-dependent probabilities to change over time.) In the case of multinomial–HMMs these probabilities are supplied by m multinomial distributions.

We note that it is not required that each of the q component series have a distribution of the same type. For example, in the bivariate model discussed in Section 11.5, the state-dependent distributions of X_{t1} are gamma distributions, and those of X_{t2} von Mises distributions; the first component is linear-valued and the second circular-valued. Secondly, it is not assumed that the m state-dependent distributions of any one series belong to the same family of distributions. In principle one could use a gamma distribution in state 1, an extreme-value distribution in state 2, and so on. However, we have not yet encountered applications in which this feature could be usefully exploited.

What is necessary is to specify models for m joint distributions, a task that can be anything but trivial. For example there is no single bivariate Poisson distribution; different versions are available and they have different properties. One has to select a version that is appropriate in the context of the particular application being investigated. (In contrast, one can reasonably speak of *the* bivariate normal distribution because, for many or most practical purposes, there is only one.)

Once the required joint distributions have been selected, i.e. once one has specified the state-dependent probabilities $_tp_i(\mathbf{x}_t)$, the likelihood of a general multivariate HMM is easy to write down. It has the same form as that of the basic model, namely

$$L_T = \boldsymbol{\delta}\,_1\mathbf{P}(\mathbf{x}_1)\boldsymbol{\Gamma}\,_2\mathbf{P}(\mathbf{x}_2)\cdots\boldsymbol{\Gamma}\,_T\mathbf{P}(\mathbf{x}_T)\mathbf{1}',$$

where $\mathbf{x}_1,\ldots,\mathbf{x}_T$ are the observations and, as before,

$$_t\mathbf{P}(\mathbf{x}_t) = \text{diag}\left(_tp_1(\mathbf{x}_t),\ldots,\,_tp_m(\mathbf{x}_t)\right).$$

The above expression for the likelihood also holds if some of the series are continuous-valued, provided that where necessary probabilities are interpreted as densities.

The task of finding suitable joint distributions is greatly simplified if one can assume **contemporaneous conditional independence**. We illustrate the meaning of this term by means of the multisite precipitation model discussed by Zucchini and Guttorp (1991). In their work there are five binary time series representing the presence or absence of rain at each of five sites which are regarded as being linked by a common weather process $\{C_t\}$. There the random variables X_{tj} are binary. Let $_t\pi_{ij}$ be defined as

$$_t\pi_{ij} = \Pr(X_{tj} = 1 \mid C_t = i) = 1 - \Pr(X_{tj} = 0 \mid C_t = i).$$

The assumption of contemporaneous conditional independence is that the state-dependent joint probability $_tp_i(\mathbf{x}_t)$ is just the product of the corresponding marginal probabilities:

$$_tp_i(\mathbf{x}_t) = \prod_{j=1}^{q} {}_t\pi_{ij}^{x_{tj}}\left(1 - {}_t\pi_{ij}\right)^{1-x_{tj}}. \tag{8.2}$$

Thus, for example, given weather state i, the probability that on day t it will rain at sites 1, 2, and 4, but not at sites 3 and 5, is the product

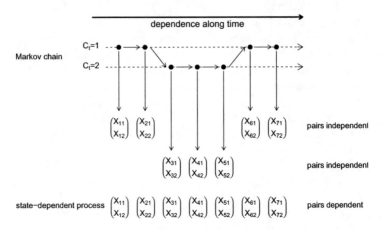

Figure 8.2 *Contemporaneous conditional independence.*

of the (marginal) probabilities that these events occur: $_t\pi_{i1}\, _t\pi_{i2}(1 - {}_t\pi_{i3})\, _t\pi_{i4}(1 - {}_t\pi_{i5})$.

For general multivariate HMMs that are contemporaneously conditionally independent, the state-dependent probabilities are given by a product of the corresponding q marginal probabilities:

$$_tp_i(\mathbf{x}_t) = \prod_{j=1}^{q} \Pr(X_{tj} = x_{tj} \mid C_t = i).$$

We wish to emphasize that the above two conditional independence assumptions, namely longitudinal conditional independence and contemporaneous conditional independence, do not imply that

- the individual component series are serially independent; or that

- the component series are mutually independent.

The parameter process, namely the Markov chain, induces both serial dependence and cross-dependence in the component series, even when these are assumed to have both of the above conditional independence properties. This is illustrated in Figure 8.2.

Details of the serial- and cross-correlation functions of these models are given in Section 3.4 of MacDonald and Zucchini (1997), as are other general classes of models for multivariate HMMs, such as models with time lags and multivariate models in which some of the components are discrete and others continuous. See also Exercise 8.

Multivariate HMMs (with continuous state-dependent distributions)

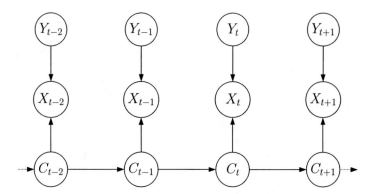

Figure 8.3 *Structure of HMM with covariates Y_t in the state-dependent probabilities.*

might for instance be used for modelling multivariate financial time series. For example, one could fit a two-state multivariate normal HMM to a multivariate series of daily returns on a number of shares where, as in the univariate case, the states of the Markov chain might correspond to calm and turbulent phases of the stock market; see Section 13.2 for such a model.

8.5 Series that depend on covariates

We now discuss two ways in which covariates can be introduced into HMMs: via the state-dependent probabilities, and via the transition probabilities of the Markov chain.

8.5.1 Covariates in the state-dependent distributions

HMMs can be modified to allow for the influence of covariates by postulating dependence of the state-dependent probabilities $_tp_i(x_t)$ on those covariates, as demonstrated in Figure 8.3. This opens the way for such models to incorporate time trend and seasonality, for instance.

Here we take $\{C_t\}$ to be the usual Markov chain, and suppose, in the case of Poisson–HMMs, that the conditional mean $_t\lambda_i = \mathrm{E}(X_t \mid C_t = i)$ depends on the (row) vector $\mathbf{y}_t$ of q covariates, for instance as follows:

$$\log \,_t\lambda_i = \beta_i\mathbf{y}_t'.$$

In the case of binomial–HMMs, the corresponding assumption is that

$$\text{logit } {}_t p_i = \beta_i \mathbf{y}'_t.$$

The elements of $\mathbf{y}_t$ could include a constant, time (t), sinusoidal components expressing seasonality (for example $\cos(2\pi t/r)$ and $\sin(2\pi t/r)$ for some positive integer r), and any other relevant covariates. For example, a binomial–HMM with

$$\text{logit } {}_t p_i = \beta_{i1} + \beta_{i2} t + \beta_{i3} \cos(2\pi t/r) + \beta_{i4} \sin(2\pi t/r) + \beta_{i5} z_t + \beta_{i6} w_t$$

allows for a (logit-linear) time trend, r-period seasonality and the influence of covariates z_t and w_t, in the state-dependent 'success probabilities' ${}_t p_i$. Additional sine-cosine pairs can if necessary be included, to model more complex seasonal patterns. Similar models for the log of the conditional mean ${}_t \lambda_i$ are possible in the Poisson–HMM case. Clearly link functions other than the canonical ones used here could instead be used. The expression for the likelihood of T consecutive observations $x_1, \ldots, x_T$ for such a model involving covariates is similar to that of the basic model:

$$L_T = \boldsymbol{\delta} \, {}_1\mathbf{P}(x_1, y_1) \boldsymbol{\Gamma}_2 \mathbf{P}(x_2, y_2) \cdots \boldsymbol{\Gamma} \, {}_T\mathbf{P}(x_T, y_T)\mathbf{1}',$$

the only difference being the allowance for covariates y_t in the state-dependent probabilities ${}_t p_i(x_t, y_t)$ and in the corresponding matrices

$$\, {}_t\mathbf{P}(x_t, y_t) = \text{diag}({}_t p_1(x_t, y_t), \ldots, {}_t p_m(x_t, y_t)).$$

Examples of models that incorporate a time trend can be found in Sections 13.1.1, 14.2 and 15.2. In Section 14.3 there are models incorporating both a time trend and a seasonal component.

It is worth noting that the binomial– and Poisson–HMMs which allow for covariates in this way provide important generalizations of logistic regression and Poisson regression respectively, generalizations that drop the independence assumption of such regression models and allow serial dependence.

8.5.2 Covariates in the transition probabilities

An alternative way of modelling time trend and seasonality in HMMs is to drop the assumption that the Markov chain is homogeneous, and assume instead that the transition probabilities are functions of time, denoted for two states as follows:

$$\, {}_t\boldsymbol{\Gamma} = \begin{pmatrix} {}_t\gamma_{11} & {}_t\gamma_{12} \\ {}_t\gamma_{21} & {}_t\gamma_{22} \end{pmatrix}.$$

More generally, the transition probabilities can be modelled as depending on one or more covariates, not necessarily time but any variables considered relevant.

Incorporation of covariates into the Markov chain is not as straightforward as incorporating them into the state-dependent probabilities. One reason why it could nevertheless be worthwhile is that the resulting Markov chain may have a useful substantive interpretation, e.g. as a weather process which is itself complex but determines rainfall probabilities at several sites in fairly simple fashion. We illustrate one way in which it is possible to modify the transition probabilities of the Markov chain in order to represent time trend and seasonality.

Consider a model based on a two-state Markov chain $\{C_t\}$ with

$$\Pr(C_t = 2 \mid C_{t-1} = 1) = {}_t\gamma_{12}, \quad \Pr(C_t = 1 \mid C_{t-1} = 2) = {}_t\gamma_{21},$$

and, for $i = 1, 2$,

$$\text{logit} \, {}_t\gamma_{i,3-i} = \boldsymbol{\beta}_i \mathbf{y}_t'.$$

For example, a model incorporating r-period seasonality is that with

$$\text{logit} \, {}_t\gamma_{i,3-i} = \beta_{i1} + \beta_{i2} \cos(2\pi t/r) + \beta_{i3} \sin(2\pi t/r).$$

In general the above assumption on $\text{logit} \, {}_t\gamma_{i,3-i}$ implies that the transition probability matrix, for transitions between times $t-1$ and t, is given by

$$
{}_t\boldsymbol{\Gamma} = \begin{pmatrix} \dfrac{1}{1 + \exp(\boldsymbol{\beta}_1 \mathbf{y}_t')} & \dfrac{\exp(\boldsymbol{\beta}_1 \mathbf{y}_t')}{1 + \exp(\boldsymbol{\beta}_1 \mathbf{y}_t')} \\ \dfrac{\exp(\boldsymbol{\beta}_2 \mathbf{y}_t')}{1 + \exp(\boldsymbol{\beta}_2 \mathbf{y}_t')} & \dfrac{1}{1 + \exp(\boldsymbol{\beta}_2 \mathbf{y}_t')} \end{pmatrix}.
$$

Extension of this model to the case $m > 2$ presents some difficulties, but they are not insuperable.

One important difference between the class of models proposed here and other HMMs (and a consequence of the nonhomogeneity of the Markov chain) is that we cannot always assume that there is a stationary distribution for the Markov chain. This problem arises when one or more of the covariates are functions of time, as in models with trend or seasonality. If necessary we therefore assume instead some initial distribution $\boldsymbol{\delta}$, i.e. a distribution for C_1.

A very general class of models in which the Markov chain is nonhomogeneous and which allows for the influence of covariates is that of Hughes (1993). This model, and additional details relating to the models outlined above, are discussed in Chapter 3 of MacDonald and Zucchini (1997).

8.6 Models with additional dependencies

In Section 8.3 we described a class of models which have dependencies in addition to those found in the basic HMM as depicted by Figure 2.2 on p. 30: second-order HMMs. In that case the additional dependencies are entirely at latent process level. Here we briefly describe three further classes of models with additional dependencies.

In the basic model of Figure 2.2 there are no edges directly connecting earlier observations to X_t; the only dependence between observations arises from the latent process $\{C_t\}$. There may well be applications, however, where extra dependencies at observation level are suspected and should be allowed for in the model. The additional computational features of such models are that the 'state-dependent' probabilities needed for the likelihood computation depend on previous observations as well as on the current state, and that the likelihood maximized is conveniently taken to be that which is conditional on the first few observations. Figure 8.4 depicts two models with such extra dependencies at observation level. In Section 14.3 we present some examples of models with additional dependencies at observation level; the likelihood evaluation and maximization proceed with little additional complication.

Some models with additional dependencies at observation level that have appeared in the literature are the double-chain Markov model of Berchtold (1999), the M1–Mk models of Nicolas *et al.* (2002) for DNA sequences, those of Boys and Henderson (2004), and hidden Markov AR(k) models, usually termed Markov-switching autoregressions.

There is another type of extra dependency which may be useful. In the basic model, the distribution of an observation depends on the current state only. It is not difficult, however, to make the minor generalization needed for that distribution to depend also on the previous state. Figure 8.5 depicts the resulting model. If we define the $m \times m$ matrix $\mathbf{Q}(x)$ to have as its (i,j) element the product $\gamma_{ij} \Pr(X_t = x \mid C_{t-1} = i, C_t = j)$, and denote by $\boldsymbol{\delta}$ the distribution of C_1, the likelihood is given by

$$L_T = \boldsymbol{\delta}\mathbf{P}(x_1)\mathbf{Q}(x_2)\mathbf{Q}(x_3)\cdots\mathbf{Q}(x_T)\mathbf{1}'. \tag{8.3}$$

If instead we denote by $\boldsymbol{\delta}$ the distribution of C_0 (not C_1), the result is somewhat neater:

$$L_T = \boldsymbol{\delta}\mathbf{Q}(x_1)\mathbf{Q}(x_2)\cdots\mathbf{Q}(x_T)\mathbf{1}'. \tag{8.4}$$

Extra dependencies of yet another kind appear in Section 13.3.3 and Chapter 16, which present respectively a discrete state-space stochastic volatility model with leverage (see Figure 13.4), and a model for animal behaviour which incorporates feedback from observation level to motivational state (see Figure 16.2). In both of these applications the extra dependencies are from observation level to the latent process.

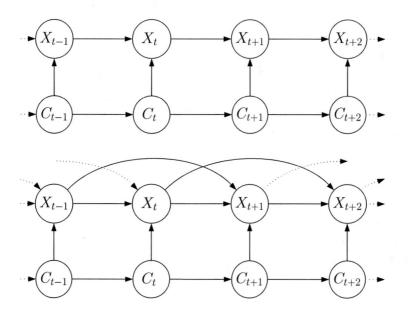

Figure 8.4 *Models with additional dependencies at observation level; the upper graph represents (e.g.) a Markov-switching AR(1) model, and the lower one a Markov-switching AR(2).*

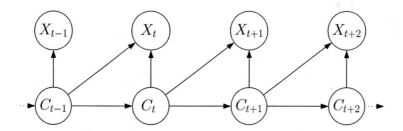

Figure 8.5 *Additional dependencies from latent process to observation level.*

Exercises

1. Give an expression for the likelihood function for each of the following HMMs. The list below refers to the state-dependent distributions.

   (a) Geometric with parameters $\theta_i \in (0,\ 1)$, $i = 1, 2, \ldots, m$.

   (b) Exponential with parameters $\lambda_i \geq 0$, $i = 1, 2, \ldots, m$.

(c) Bivariate normal with parameters

$$\boldsymbol{\mu}_i = \begin{pmatrix} \mu_{1i} \\ \mu_{2i} \end{pmatrix}, \qquad \boldsymbol{\Sigma}_i = \begin{pmatrix} \sigma_{1i}^2 & \sigma_{12i} \\ \sigma_{12i} & \sigma_{2i}^2 \end{pmatrix},$$

for $i = 1, 2, \ldots, m$.

2. Let $\{X_t\}$ be a second-order HMM, based on a stationary second-order Markov chain $\{C_t\}$ on m states. How many parameters does the model have in the following cases? (Assume for simplicity that one parameter is needed to specify each of the m state-dependent distributions.)

(a) $\{C_t\}$ is a general second-order Markov chain.

(b) $\{C_t\}$ is a Pegram model.

(c) $\{C_t\}$ is a Raftery model.

3. Let $\{X_t\}$ be a second-order HMM, based on a stationary second-order Markov chain $\{C_t\}$ on m states. For integers $t \geq 2$, and integers i and j from 1 to m, define

$$\nu_t(i, j; \mathbf{x}^{(t)}) = \Pr(\mathbf{X}^{(t)} = \mathbf{x}^{(t)}, C_{t-1} = i, C_t = j).$$

Note that these probabilities are just an extension to two dimensions of the forward probabilities $\alpha_t(i)$, which could more explicitly be denoted by $\alpha_t(i; \mathbf{x}^{(t)})$. For the case $t = 2$ we have

$$\nu_2(i, j; \mathbf{x}^{(t)}) = u(i, j)p_i(x_1)p_j(x_2).$$

(a) Show that, for integers $t \geq 3$,

$$\nu_t(j, k; \mathbf{x}^{(t)}) = \left(\sum_{i=1}^m \nu_{t-1}(i, j; \mathbf{x}^{(t-1)}) \, \gamma(i, j, k) \right) p_k(x_t). \qquad (8.5)$$

(b) Show how the recursion (8.5) can be used to compute the likelihood of a series of T observations, $\Pr(\mathbf{X}^{(T)} = \mathbf{x}^{(T)})$.

(c) Show that the computational effort required to find the likelihood thus is $O(Tm^3)$.

4. Consider the generalization of the basic HMM which allows the distribution of an observation to depend on the current state of a Markov chain *and* the previous one; see Section 8.6. Prove the results given in Equations (8.3) and (8.4) for the likelihood of such a model.

5. Find the autocorrelation functions for stationary HMMs with (a) normal and (b) binomial state-dependent distributions.

6. (Runlengths in Bernoulli–HMMs) Let $\{X_t\}$ be a Bernoulli–HMM in which the underlying stationary irreducible Markov chain has the states 1, 2, ..., m, transition probability matrix $\boldsymbol{\Gamma}$ and stationary

distribution $\boldsymbol{\delta}$. The probability of an observation being 1 in state i is denoted by p_i. Define a run of ones as follows: such a run is initiated by the sequence 01, and is said to be of length $k \in \mathbb{N}$ if that sequence is followed by a further $k - 1$ ones and a zero (in that order).

(a) Let K denote the length of a run of ones. Show that

$$\Pr(K=k) = \frac{\boldsymbol{\delta}\mathbf{P}(0)(\boldsymbol{\Gamma}\mathbf{P}(1))^k\boldsymbol{\Gamma}\mathbf{P}(0)\mathbf{1}'}{\boldsymbol{\delta}\mathbf{P}(0)\boldsymbol{\Gamma}\mathbf{P}(1)\mathbf{1}'},$$

where $\mathbf{P}(1) = \mathrm{diag}(p_1, \ldots, p_m)$ and $\mathbf{P}(0) = \mathbf{I}_m - \mathbf{P}(1)$.

(b) Suppose that $\mathbf{B} = \boldsymbol{\Gamma}\mathbf{P}(1)$ has distinct eigenvalues w_i. Show that the probability $\Pr(K=k)$ is, as a function of k, a linear combination of the kth powers of these eigenvalues, and hence of $w_i^{k-1}(1 - w_i)$, $i=1,\ldots,m$.

(c) Does this imply that K is a mixture of geometric random variables? (Hint: will the eigenvalues w_i always lie between zero and one?)

(d) Assume for the rest of this exercise that there are only two states, with the t.p.m. given by

$$\boldsymbol{\Gamma} = \begin{pmatrix} 1 - \gamma_{12} & \gamma_{12} \\ \gamma_{21} & 1 - \gamma_{21} \end{pmatrix},$$

and that $p_1 = 0$ and $p_2 \in (0, 1)$.

Show that the distribution of K is as follows, for all $k \in \mathbb{N}$:

$$\Pr(K=k) = ((1 - \gamma_{21})p_2)^{k-1}(1 - (1 - \gamma_{21})p_2).$$

(So although not itself a Markov chain, this HMM has a geometric distribution for the length of a run of ones.)

(e) For such a model, will the length of a run of zeros also be geometrically distributed?

7. Consider the three two-state stationary Bernoulli–HMMs specified below. For instance, in model (a), $\Pr(X_t = 1 \mid C_t = 1) = 0.1$ and $\Pr(X_t = 1 \mid C_t = 2) = 1$, and X_t is either one or zero. The states are determined in accordance with the stationary Markov chain with t.p.m. $\boldsymbol{\Gamma}$. (Actually, (c) is a Markov chain; there is in that case a one-to-one correspondence between states and observations.)

Let K denote the length of a run of ones. In each of the three cases, determine the following: $\Pr(K = k)$, $\Pr(K \leq 10)$, $\mathrm{E}(K)$, σ_K and $\mathrm{corr}(X_t, X_{t+k})$.

(a)

$$\boldsymbol{\Gamma} = \begin{pmatrix} 0.99 & 0.01 \\ 0.08 & 0.92 \end{pmatrix}, \qquad \mathbf{p} = (0.1, 1).$$

(b)
$$\Gamma = \begin{pmatrix} 0.98 & 0.02 \\ 0.07 & 0.93 \end{pmatrix}, \qquad \mathbf{p} = (0, 0.9).$$

(c)
$$\Gamma = \begin{pmatrix} 0.9 & 0.1 \\ 0.4 & 0.6 \end{pmatrix}, \qquad \mathbf{p} = (0, 1).$$

Notice that these three models are comparable in that (i) the unconditional probability of an observation being one is in all cases 0.2; and
(ii) all autocorrelations are positive and decrease geometrically.

8. Consider a two-state bivariate HMM $\{(X_{t1}, X_{t2}) : t \in \mathbb{N}\}$, based on
a stationary Markov chain with transition probability matrix Γ and
stationary distribution $\boldsymbol{\delta} = (\delta_1, \delta_2)$. Let μ_{i1} and μ_{i2} denote the means
of X_{t1} and X_{t2} in state i, and similarly σ_{i1}^2 and σ_{i2}^2 the variances.
Assume both contemporaneous conditional independence and longi-
tudinal conditional independence.

(a) Show that, for all nonnegative integers k,

$$\mathrm{Cov}(X_{t1}, X_{t+k,2}) = \delta_1 \delta_2 (\mu_{11} - \mu_{21})(\mu_{12} - \mu_{22})(1 - \gamma_{12} - \gamma_{21})^k.$$

(b) Hence find the cross-correlations $\mathrm{Corr}(X_{t1}, X_{t+k,2})$.

(c) Does contemporaneous conditional independence imply independence of X_{t1} and X_{t2}?

(d) State a sufficient condition for X_{t1} and X_{t2} to be uncorrelated.

(e) Generalize the results of (a) and (b) to any number (m) of states.

9. (Irreversibility of a binomial–HMM) In Section 1.3.3 we defined reversibility for a random process, and showed that the stationary Markov chain with the t.p.m. Γ given below is not reversible.

$$\Gamma = \begin{pmatrix} 1/3 & 1/3 & 1/3 \\ 2/3 & 0 & 1/3 \\ 1/2 & 1/2 & 0 \end{pmatrix}.$$

Let $\{X_t\}$ be the stationary HMM with Γ as above, and having binomial state-dependent distributions with parameters 2 and 0/0.5/1;
e.g. in state 1 the observation X_t is distributed Binomial(2, 0).

By finding the probabilities $\Pr(X_t = 0, X_{t+1} = 1)$ and $\Pr(X_t = 1, X_{t+1} = 0)$, or otherwise, show that $\{X_t\}$ is irreversible.

PART TWO

Applications

CHAPTER 9

Epileptic seizures

9.1 Introduction

Albert (1991) and Le, Leroux and Puterman (1992) describe the fitting
of two-state Poisson–HMMs to series of daily counts of epileptic seizures
in one patient. Such models appear to be a promising tool for the analysis
of seizure counts, the more so as there are suggestions in the neurology
literature that the susceptibility of a patient to seizures may vary in
a fashion that can reasonably be represented by a Markov chain; see
Hopkins, Davies and Dobson (1985). Another promising approach, not
pursued here, is to use an AR(1) analogue based on thinning; see Franke
and Seligmann (1993).

9.2 Models fitted

Table 9.1 *Counts of epileptic seizures in one patient on 204 consecutive days
(to be read across rows).*

0 3 0 0 0 0 1	1 0 2 1 1 2 0	0 1 2 1 3 1 3
0 4 2 0 1 1 2	1 2 1 1 1 0 1	0 2 2 1 2 1 0
0 0 2 1 2 0 1	0 1 0 1 0 0 0	0 0 0 0 0 1 0
0 0 0 0 1 0 0	0 1 0 0 0 1 0	0 0 1 0 0 1 0
0 2 1 0 1 1 0	0 0 2 2 0 1 1	3 1 1 2 1 0 3
6 1 3 1 2 2 1	0 1 2 1 0 1 2	0 0 2 2 1 0 1
0 0 2 0 1 0 0	0 1 0 0 1 0 0	0 0 0 0 0 1 3
0 0 0 0 0 1 0	1 1 1 0 0 0 0	0 1 0 1 2 1 0
0 0 0 0 0 1 4	0 0 0 0 0 0 0	0 0 0 0 0 0 0
0 0 0 0 0 0 0	0 0 0 0 0 0 0	0

We analyse here a series of counts of myoclonic seizures suffered by one
patient on 204 consecutive days*. The observations are given in Table
9.1 and displayed in Figure 9.1.

---

* The 225-day series published by Le *et al.* (1992) contained a repeat of the observations for a 21-day period; see MacDonald and Zucchini (1997, p. 208). Table 9.1
gives the corrected series.

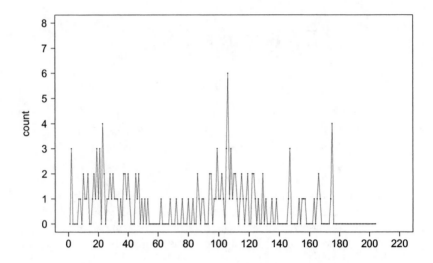

Figure 9.1 *Epileptic seizure counts on 204 days.*

Le *et al.* use an HMM of the type described by Leroux and Puterman (1992). Their model does not assume that the underlying Markov chain is stationary. It is fitted by maximizing the likelihood conditional on the Markov chain starting in a given state with probability one, and then maximizing over the possible initial states.

We consider a similar HMM, but based on a stationary Markov chain and fitted by maximization of the unconditional likelihood of the observations. We investigate models with $m = 1$, 2, 3 and 4 states. (The one-state model is just the model which assumes that the observations are realizations of independent Poisson random variables with a common mean. That mean is the only parameter.)

Table 9.2 gives the AIC and BIC values for the models. From the table we see that, of the four models considered, the three-state model is chosen by AIC, but the two-state model is chosen, by a large margin, by BIC. We concentrate on the two-state model, the details of which are as follows. The Markov chain has transition probability matrix

$$\begin{pmatrix} 0.965 & 0.035 \\ 0.027 & 0.973 \end{pmatrix},$$

Table 9.2 *Epileptic seizure counts: comparison of several stationary Poisson–HMMs by means of AIC and BIC.*

no. of states	k	$-l$	AIC	BIC
1	1	232.15	466.31	469.63
2	4	211.68	431.36	**444.64**
3	9	205.55	**429.10**	458.97
4	16	201.68	435.36	488.45

Table 9.3 *Sample ACF for the epileptic seizure counts.*

k	1	2	3	4	5	6	7	8
$\hat{\rho}(k)$	0.236	0.201	0.199	0.250	0.157	0.181	0.230	0.242

and starts from the stationary distribution (0.433, 0.567). The seizure rates in states 1 and 2 are 1.167 and 0.262 respectively.

The ACF of the model can be computed by the results of Exercise 4 of Chapter 2. It is given, for all positive integers k, by

$$
\rho(k) = \left(1 + \frac{\delta\lambda'}{(\lambda_2 - \lambda_1)^2\delta_1\delta_2}\right)^{-1}(1 - \gamma_{12} - \gamma_{21})^k
$$
$$
= 0.235 \times 0.939^k.
$$

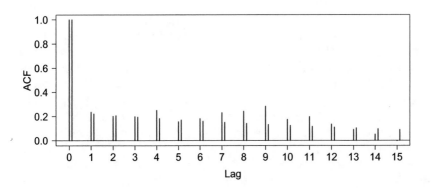

Figure 9.2 *Sample and theoretical ACF for the epileptic seizures data. At each lag the left bar represents the sample ACF, and the right bar the ACF of a stationary two-state Poisson–HMM.*

Table 9.4 *Observed and expected numbers of days with $r = 0, 1, 2, \ldots$ epileptic seizures.*

r	observed no.	expected no.
0	117	116.5
1	54	55.4
2	23	21.8
3	7	7.5
4	2	2.1
5	0	0.5
≥ 6	1	0.1
	204	203.9

Table 9.3 gives the corresponding sample ACF. Figure 9.2, which displays both, shows that the agreement between sample and theoretical ACF is reasonably close.

The marginal properties of the model can be assessed from Table 9.4, which gives the observed and expected numbers of days on which there were 0, 1, 2, ..., 6 or more seizures. Agreement is excellent.

9.3 Model checking by pseudo-residuals

We now use the techniques of Section 6.2.2 to check for outliers under the two-state model we have chosen. Figure 9.3 is a plot of ordinary normal pseudo-residual segments. From it we see that three observations of the 204 stand out as extreme, namely those for days 106, 147 and 175. They all yield pseudo-residual segments lying entirely within the top $\frac{1}{2}\%$ of their respective distributions.

It is interesting to note that observations 23 and 175, both of which represent four seizures in one day, yield rather different pseudo-residual segments. The reason for this is clear when one notes that most of the near neighbours (in time) of observation 175 are zero, which is not true of observation 23 and its neighbours. Observation 23 is much less extreme relative to its neighbours than is 175, and this is reflected in the pseudo-residual. Similarly, observation 106 (six seizures in a day) is less extreme relative to its neighbours than is observation 175 (four seizures).

However, a more interesting exercise is to see whether, if a model had been fitted to (say) the first 100 observations only, day-by-day monitoring thereafter by means of forecast pseudo-residuals would have identified any outliers. The two-state model fitted from the first 100 observa-

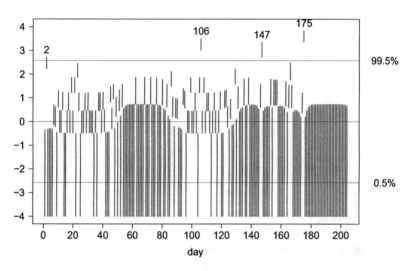

Figure 9.3 *Epileptic seizures data: ordinary (normal) pseudo-residual segments, relative to stationary two-state Poisson–HMM fitted to all 204 observations.*

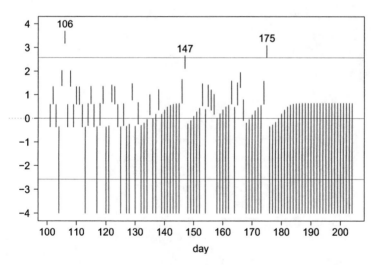

Figure 9.4 *Epileptic seizures data: forecast pseudo-residual segments, relative to stationary two-state Poisson–HMM fitted to data for days 1–100 only.*

tions has transition probability matrix

$$\begin{pmatrix} 0.983 & 0.017 \\ 0.042 & 0.958 \end{pmatrix},$$

and seizure rates 1.049 and 0.258.

From a plot of forecast pseudo-residuals, Figure 9.4, we see that the same three observations stand out: days 106, 147 and 175. Observation 106 emerges from such a monitoring procedure as the clearest outlier relative to its predecessors, then 175, then 147.

Exercises

1. Consider the two-state model for the epileptic seizures.

   (a) Compute the probabilities $\Pr(C_t = i \mid \mathbf{X}^{(T)})$ for this model, for $i = 1, 2$ and all t.

   (b) Perform both local and global decoding to estimate the most likely states. Do the results differ?

   (c) Perform state prediction for the next three time points; i.e. find the probabilities $\Pr(C_{T+h} = i \mid \mathbf{X}^{(T)})$ for $h = 1, 2, 3$.

   (d) Compute the forecast distribution $\Pr(X_{T+h} = x \mid \mathbf{X}^{(T)})$ for $h = 1$, ..., 10.

2.(a) Fit a stationary three-state Poisson–HMM to the epileptic seizures.

   (b) Find the general expression for the ACF of this model.

   (c) For lags 1 to 8, compare this model ACF with the sample ACF given in Table 9.3.

CHAPTER 10

Eruptions of the Old Faithful geyser

10.1 Introduction

There are many published analyses, from various points of view, of data relating to eruptions of the Old Faithful geyser in the Yellowstone National Park in the USA: for instance Cook and Weisberg (1982, pp. 40–42), Weisberg (1985, pp. 230–235), Silverman (1985; 1986, p. 7), Scott (1992, p. 278), and Aston and Martin (2007). Some of these accounts ignore the strong serial dependence in the behaviour of the geyser; see the comments of Diggle (1993).

In this chapter we present:

- an analysis of a series of long and short eruption durations of the geyser. This series is a dichotomized version of one of the two series provided by Azzalini and Bowman (1990).

- univariate models for the series of durations and waiting times, in their original, non-dichotomized, form; and

- a bivariate model for the durations and waiting times.

The models we describe are mostly HMMs, but in Section 10.2 we also fit Markov chains of first and second order, and compare them with the HMMs.

10.2 The binary time series of short and long eruptions

Azzalini and Bowman (1990) have presented a time series analysis of data on eruptions of Old Faithful. The data consist of 299 pairs of observations, collected continuously from 1 August to 15 August 1985. The pairs are (w_t, d_t), with w_t being the time between the starts of successive eruptions, and d_t being the duration of the subsequent eruption.

It is true of both series that most of the observations can be described as either long or short, with very few observations intermediate in length, and with relatively low variation within the low and high groups. It is therefore natural to treat these series as binary time series; Azzalini and Bowman do so by dichotomizing the 'waiting times' w_t at 68 minutes and the durations d_t at 3 minutes, denoting short by 0 and long by 1. There is, in respect of the durations series, the complication that some of the eruptions were recorded only as short, medium or long, and the medium durations have to be treated as either short or long. We use the convention that the (two) mediums are treated as long*.

Table 10.1 *Short and long eruption durations of Old Faithful geyser (299 observations, to be read across rows).*

1 0 1 1 1 0 1 1 0 1 0 1 0 1 1 0 1 0 1 1 0 1 0 1 0 1 0 1 1 1
1 1 0 1 0 1 0 1 0 1 0 1 0 1 0 1 0 1 0 1 0 1 0 1 0 1 1 1 1 1
0 1 0 1 0 1 0 1 1 0 1 0 1 1 1 0 1 1 1 1 0 1 1 1 0 1 0 1 0
1 0 1 0 1 0 1 0 1 0 1 0 1 0 1 0 1 0 1 0 1 1 0 1 0 1 0 1 0 1
0 1 1 1 0 1 1 1 1 1 1 0 1 1 1 1 0 1 1 1 1 1 1 0 1 0 1
0 1 0 1 0 1 0 1 1 1 1 1 0 1 0 1 0 1 0 1 1 1 0 1 0 1 0 1 1
0 1 0 1 1 1 1 0 1 0 1 0 1 0 1 1 1 0 1 0 1 0 1 1 0 1 1 0 1 1
1 0 1 0 1 0 1 0 1 1 0 1 1 1 1 1 1 0 1 0 1 0 1 1 1 1 0 1 1
0 1 1 1 0 1 1 0 1 0 1 1 1 0 1 0 1 1 1 1 1 0 1 1 1 0 1 0 1 0
1 1 0 1 0 1 1 1 1 1 1 1 1 0 1 0 1 0 1 0 1 0 1 0 1 0 1 1 0

It emerges that $\{W_t\}$ and $\{D_t\}$, the dichotomized versions of the series $\{w_t\}$ and $\{d_t\}$, are very similar — almost identical, in fact — and Azzalini and Bowman therefore concentrate on the series $\{D_t\}$ as representing most of the information relevant to the state of the system. Table 10.1 presents the series $\{D_t\}$. On examination of this series one notices that 0 is always followed by 1, and 1 by either 0 or 1. A summary of the data is displayed in the 'observed no.' column of Table 10.3 on p. 144.

10.2.1 Markov chain models

What Azzalini and Bowman first did was to fit a (first-order) Markov chain model. This model seemed quite plausible from a geophysical point of view, but did not match the sample ACF at all well. They then fitted a second-order Markov chain model, which matched the ACF much better, but did not attempt a geophysical interpretation for this second model. We describe what Azzalini and Bowman did, and then fit HMMs and compare them with their models.

Using the estimator of the ACF described by Box, Jenkins and Reinsel (1994, p. 31), Azzalini and Bowman estimated the ACF and PACF of $\{D_t\}$ as displayed in Table 10.2. Since the sample ACF is not even

---

* Azzalini and Bowman appear to have used one convention when estimating the ACF (medium=long), and the other when estimating the t.p.m. (medium=short). There are only very minor differences between their results and ours.

Table 10.2 *Old Faithful eruptions: sample autocorrelation function ($\hat{\rho}(k)$) and partial autocorrelation function ($\hat{\phi}_{kk}$) of the series $\{D_t\}$ of short and long eruptions.*

k	1	2	3	4	5	6	7	8
$\hat{\rho}(k)$	−0.538	0.478	−0.346	0.318	−0.256	0.208	−0.161	0.136
$\hat{\phi}_{kk}$	−0.538	0.266	−0.021	0.075	−0.021	−0.009	0.010	0.006

approximately of the form α^k, a Markov chain is not a satisfactory model; see Section 1.3.4. Azzalini and Bowman therefore fitted a second-order Markov chain, which turned out not to be consistent with a first-order model. They mention also that they fitted a third-order model, which did produce estimates consistent with a second-order model.

An estimate of the transition probability matrix of the first-order Markov chain, based on maximizing the likelihood conditional on the first observation as described in Section 1.3.5, is

$$\begin{pmatrix} 0 & 1 \\ \frac{105}{194} & \frac{89}{194} \end{pmatrix} = \begin{pmatrix} 0 & 1 \\ 0.5412 & 0.4588 \end{pmatrix}. \tag{10.1}$$

Although it is not central to this discussion, it is worth noting that unconditional maximum likelihood estimation is very easy in this case. Because there are no transitions from 0 to 0, the explicit result of Bisgaard and Travis (1991) applies; see our Equation (1.6) on p. 22. The result is that the transition probability matrix is estimated as

$$\begin{pmatrix} 0 & 1 \\ 0.5404 & 0.4596 \end{pmatrix}. \tag{10.2}$$

This serves to confirm as reasonable the expectation that, for a series of length 299, estimation by conditional maximum likelihood differs very little from unconditional.

Since the sequence $(0,0)$ does not occur, the three states needed to express the second-order Markov chain as a first-order Markov chain are, in order: $(0,1)$, $(1,0)$, $(1,1)$. The corresponding t.p.m. is

$$\begin{pmatrix} 0 & \frac{69}{104} & \frac{35}{104} \\ 1 & 0 & 0 \\ 0 & \frac{35}{89} & \frac{54}{89} \end{pmatrix} = \begin{pmatrix} 0 & 0.6635 & 0.3365 \\ 1 & 0 & 0 \\ 0 & 0.3933 & 0.6067 \end{pmatrix}. \tag{10.3}$$

The model (10.3) has stationary distribution $\frac{1}{297}(104, 104, 89)$, and the

Table 10.3 *Old Faithful: observed numbers of short and long eruptions and various transitions, compared with those expected under the two-state HMM.*

	observed no.	expected no.
short eruptions (0)	105	105.0
long eruptions (1)	194	194.0
Transitions:		
from 0 to 0	0	0.0
from 0 to 1	104	104.0
from 1 to 0	105	104.9
from 1 to 1	89	89.1
from (0,1) to 0	69	66.7
from (0,1) to 1	35	37.3
from (1,0) to 1	104	104.0
from (1,1) to 0	35	37.6
from (1,1) to 1	54	51.4

ACF can be computed from

$$\rho(k) \quad = \quad \frac{E(D_t D_{t+k}) - E(D_t)E(D_{t+k})}{\text{Var}(D_t)}$$

$$= \quad \frac{297^2 \Pr(D_t = D_{t+k} = 1) - 193^2}{193 \times 104}.$$

The resulting figures for $\{\rho(k)\}$ are given in Table 10.4 on p. 145, and match the sample ACF $\{\hat{\rho}(k)\}$ well.

10.2.2 Hidden Markov models

We now discuss the use of HMMs for the series $\{D_t\}$. Bernoulli–HMMs with $m = 1, 2, 3$ and 4 were fitted to this series.

We describe the two-state model in some detail. This model has log-likelihood -127.31, $\mathbf{\Gamma} = \begin{pmatrix} 0.000 & 1.000 \\ 0.827 & 0.173 \end{pmatrix}$ and state-dependent probabilities of a long eruption given by the vector $(0.225, 1.000)$. That is, there are two (unobserved) states, state 1 always being followed by state 2, and state 2 by state 1 with probability 0.827. In state 1 a long eruption has probability 0.225, in state 2 it has probability 1. A convenient interpretation of this model is that it is a rather special stationary two-state Markov chain, with some noise present in the first state; if the probability 0.225 were instead zero, the model would be exactly a Markov chain. Since (in the usual notation) $\mathbf{P}(1) = \text{diag}(0.225, 1.000)$ and $\mathbf{P}(0) = \mathbf{I}_2 - \mathbf{P}(1)$, a long eruption has unconditional probability

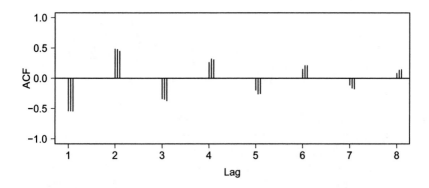

Figure 10.1 *Old Faithful, short and long eruptions: sample autocorrelation function, and ACF of two models. At each lag the centre bar represents the sample ACF, the left bar the ACF of the second-order Markov chain (i.e. model (10.3)), and the right bar that of the two-state HMM.*

Table 10.4 *Old Faithful, short and long eruptions: sample ACF compared with the ACF of the second-order Markov chain and the HMM.*

k	1	2	3	4	5	6	7	8
$\rho(k)$ for model (10.3)	−0.539	0.482	−0.335	0.262	−0.194	0.147	−0.110	0.083
sample ACF, $\hat{\rho}(k)$	−0.538	0.478	−0.346	0.318	−0.256	0.208	−0.161	0.136
$\rho(k)$ for HM model	−0.541	0.447	−0.370	0.306	−0.253	0.209	−0.173	0.143

$\Pr(X_t = 1) = \boldsymbol{\delta}\mathbf{P}(1)\mathbf{1}' = 0.649$, and a long eruption is followed by a short one with probability $\boldsymbol{\delta}\mathbf{P}(1)\boldsymbol{\Gamma}\mathbf{P}(0)\mathbf{1}'/\boldsymbol{\delta}\mathbf{P}(1)\mathbf{1}' = 0.541$. A short is always followed by a long. A comparison of observed numbers of zeros, ones and transitions, with the numbers expected under this model, is presented in Table 10.3. (A similar comparison in respect of the second-order Markov chain model would not be informative because in that case parameters have been estimated by a method that forces equality of observed and expected numbers of first- and second-order transitions.)

The ACF is given for all $k \in \mathbb{N}$ by $\rho(k) = (1 + \alpha)^{-1}w^k$, where $w = -0.827$ and $\alpha = 0.529$. Hence $\rho(k) = 0.654 \times (-0.827)^k$. In Figure 10.1 and Table 10.4 the resulting figures are compared with the sample ACF and with the theoretical ACF of the second-order Markov chain model (10.3). It seems reasonable to conclude that the HMM fits the sample

Table 10.5 *Old Faithful, short and long eruptions: percentiles of bootstrap sample of estimators of parameters of two-state HMM.*

percentile:	5th	25th	median	75th	95th
$\hat{\gamma}_{21}$	0.709	0.793	0.828	0.856	0.886
$\hat{p}_1$	0.139	0.191	0.218	0.244	0.273

ACF well; not quite as well as the second-order Markov chain model as regards the first three autocorrelations, but better for longer lags.

The parametric bootstrap, with a sample size of 100, was used to estimate the means and covariances of the maximum likelihood estimators of the four parameters γ_{12}, γ_{21}, p_1 and p_2. That is, 100 series of length 299 were generated from the two-state HMM described above, and a model of the same type fitted in the usual way to each of these series. The sample mean vector for the four parameters is (1.000, 0.819, 0.215, 1.000), and the sample covariance matrix is

$$\begin{pmatrix} 0 & 0 & 0 & 0 \\ 0 & 0.003303 & 0.001540 & 0 \\ 0 & 0.001540 & 0.002065 & 0 \\ 0 & 0 & 0 & 0 \end{pmatrix}.$$

The estimated standard deviations of the estimators are therefore (0.000, 0.057, 0.045, 0.000). (The zero standard errors are of course not typical; they are a consequence of the rather special nature of the model from which we are generating the series. Because the model has $\hat{\gamma}_{12} = 1$ and $\hat{p}_2 = 1$, the generated series have the property that a short is always followed by a long, and all the models fitted to the generated series also have $\hat{\gamma}_{12} = 1$ and $\hat{p}_2 = 1$.)

As a further indication of the behaviour of the estimators we present in Table 10.5 selected percentiles of the bootstrap sample of values of $\hat{\gamma}_{21}$ and $\hat{p}_1$. From these bootstrap results it appears that, for this application, the maximum likelihood estimators have fairly small standard deviations and are not markedly asymmetric. It should, however, be borne in mind that the estimate of the distribution of the estimators which is provided by the parametric bootstrap is derived under the assumption that the model fitted is correct.

There is a further class of models that generalizes both the two-state second-order Markov chain and the two-state HMM as described above. This is the class of two-state second-order HMMs, described in Section 8.3. By using the recursion (8.5) for the probability $\nu_t(j, k; \mathbf{x}_t)$, with the appropriate scaling, it is almost as straightforward to compute the likelihood of a second-order model as a first-order one and to fit models by maximum likelihood. In the present example the resulting probabilities of a long eruption are 0.0721 (state 1) and 1.0000 (state 2). The parameter process is a two-state second-order Markov chain with associated first-order Markov chain having transition probability matrix

$$\begin{pmatrix} 1-a & a & 0 & 0 \\ 0 & 0 & 0.7167 & 0.2833 \\ 0 & 1.0000 & 0 & 0 \\ 0 & 0 & 0.4414 & 0.5586 \end{pmatrix}. \tag{10.4}$$

Here a may be taken to be any real number between 0 and 1, and the four states used for this purpose are, in order: (1,1), (1,2), (2,1), (2,2). The log-likelihood is -126.9002. (Clearly the state (1,1) can be disregarded above without loss of information, in which case the first row and first column are deleted from the matrix (10.4).)

It should be noted that the second-order Markov chain used here as the underlying process is the general four-parameter model, not the Pegram–Raftery submodel, which has three parameters. From the comparison which follows it will be seen that an HMM based on a Pegram–Raftery second-order chain is in this case not worth pursuing, because with a total of five parameters it cannot produce a log-likelihood value better than -126.90. (The two four-parameter models fitted produce values of -127.31 and -127.12, which by AIC and BIC would be preferable to a log-likelihood of -126.90 for a five-parameter model.)

10.2.3 Comparison of models

We now compare all the models considered so far, on the basis of their unconditional log-likelihoods, denoted by l, and AIC and BIC. For instance, in the case of model (10.1), the first-order Markov chain fitted by conditional maximum likelihood, we have

$$l = \log(194/299) + 105\log(105/194) + 89\log(89/194) = -134.2426.$$

The comparable figure for model (10.2) is -134.2423; in view of the minute difference we shall here ignore the distinction between estimation by conditional and by unconditional maximum likelihood. For the second-order Markov chain model (10.3) we have

$$l = \log(104/297) + 35\log(35/104) + 69\log(69/104)$$
$$+ 35\log(35/89) + 54\log(54/89) = -127.12.$$

Table 10.6 presents a comparison of seven types of model, including for completeness the one-state HMM, i.e. the model which assumes independence of the consecutive observations. (Given the strong serial de-

Table 10.6 *Old Faithful, short and long eruptions: comparison of models on the basis of AIC and BIC.*

model	k	$-l$	AIC	BIC
1-state HM (i.e. independence)	1	193.80	389.60	393.31
Markov chain	2	134.24	272.48	279.88
second-order Markov chain	4	127.12	**262.24**	**277.04**
2-state HM	4	127.31	262.62	277.42
3-state HM	9	126.85	271.70	305.00
4-state HM	16	126.59	285.18	344.39
2-state second-order HM	6	126.90	265.80	288.00

pendence apparent in the data, it is not surprising that the one-state model is so much inferior to the others considered.)

From the table it emerges that, on the basis of AIC and BIC, only the second-order Markov chain and the two-state (first-order) HMM are worth considering. In the comparison, both of these models are taken to have four parameters, because, although the observations suggest that the sequence (short, short) cannot occur, there is no *a priori* reason to make such a restriction.

While it is true that the second-order Markov chain seems a slightly better model on the basis of the model selection exercise described above, and possibly on the basis of the ACF, both are reasonable models capable of describing the principal features of the data without using an excessive number of parameters. The HMM perhaps has the advantage of relative simplicity, given its nature as a Markov chain with some noise in one of the states. Azzalini and Bowman note that their second-order Markov chain model would require a more sophisticated interpretation than does their first-order model. Either a longer series of observations or a convincing geophysical interpretation for one model rather than the other would be needed to take the discussion further.

10.2.4 Forecast distributions

The ratio of likelihoods, as described in Section 5.2, can be used to provide the forecasts implied by the fitted two-state HMM. As it happens, the last observation in the series, D_{299}, is 0, so that under the model $\Pr(D_{300} = 1) = 1$. The conditional distribution of the next h values given the history $\mathbf{D}^{(299)}$, i.e. the joint h-step ahead forecast, is easily computed. For $h = 3$ this is given in Table 10.7. The corresponding

Table 10.7 *Old Faithful, short and long eruptions: the probabilities* $\Pr(D_{300} = 1, D_{301} = i, D_{302} = j \mid \mathbf{D}^{(299)})$ *for the two-state HMM (left) and the second-order Markov chain model (right).*

	$j = 0$	1			$j = 0$	1
$i = 0$	0.000	0.641		$i = 0$	0.000	0.663
1	0.111	0.248		1	0.132	0.204

probabilities for the second-order Markov chain model are also given in the table.

10.3 Univariate normal–HMMs for durations and waiting times

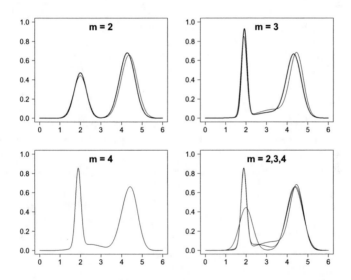

Figure 10.2 *Old Faithful durations, normal–HMMs. Thick lines (m = 2 and 3 only): models based on continuous likelihood. Thin lines (all panels): models based on discrete likelihood.*

Bearing in mind the dictum of van Belle (2002, p. 99) that one should not dichotomize unless absolutely necessary, we now describe normal–HMMs for the durations and waiting-times series in their original, non-dichotomized, form. Note that the quantities denoted w_t, both by Azzalini and Bowman (1990) and by us, are the times between the *starts*

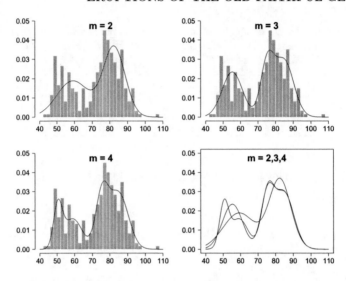

Figure 10.3 *Old Faithful waiting times, normal–HMMs. Models based on continuous likelihood and models based on discrete likelihood are essentially the same. Notice that the model for $m = 3$ is identical, or almost identical, to the three-state model of Robert and Titterington (1998): see their Figure 7.*

of successive eruptions, and each w_t therefore consists of an eruption duration plus an interval that lies strictly between eruptions. (Azzalini and Bowman note that w_t therefore exceeds the true waiting time, but ignore this because the eruption durations d_t are small relative to the quantities w_t.)

We use here the discrete likelihood, which accepts observations in the form of upper and lower bounds. Section 1.2.4 contains a description of the bounds used for the durations: see p. 13. The waiting times are all given by Azzalini and Bowman to the nearest minute, so (for instance) $w_1 = 80$ means that $w_1 \in (79.5, 80.5)$.

First we present the likelihood and AIC and BIC values of the normal–HMMs. We fitted univariate normal–HMMs with two to four states to durations and waiting times, and compared these on the basis of AIC and BIC. For both durations and waiting times, the four-state model is chosen by AIC and the three-state by BIC. We concentrate now on the three-state models, which are given in Table 10.10. In all cases the model quoted is that based on maximizing the discrete likelihood, and the states have been ordered in increasing order of mean. The transition probability matrix and stationary distribution are, as usual, $\mathbf{\Gamma}$ and $\boldsymbol{\delta}$, and the state-dependent means and standard deviations are μ_i and σ_i,

Table 10.8 *Old Faithful durations: comparison of normal–HMMs and independent mixture models by AIC and BIC, all based on discrete likelihood.*

model	k	$-\log L$	AIC	BIC
2-state HM	6	1168.955	2349.9	2372.1
3-state HM	12	1127.185	2278.4	**2322.8**
4-state HM	20	1109.147	**2258.3**	2332.3
indep. mixture (2)	5	1230.920	2471.8	2490.3
indep. mixture (3)	8	1203.872	2423.7	2453.3
indep. mixture (4)	11	1203.636	2429.3	2470.0

Table 10.9 *Old Faithful waiting times: comparison of normal–HMMs based on discrete likelihood.*

model	k	$-\log L$	AIC	BIC
2-state HM	6	1092.794	2197.6	2219.8
3-state HM	12	1051.138	2126.3	**2170.7**
4-state HM	20	1038.600	**2117.2**	2191.2

Table 10.10 *Old Faithful: three-state (univariate) normal–HMMs, based on discrete likelihood.*

Durations:

Γ				i	1	2	3
0.000	0.000	1.000		δ_i	0.291	0.195	0.514
0.053	0.113	0.834		μ_i	1.894	3.400	4.459
0.546	0.337	0.117		σ_i	0.139	0.841	0.320

Waiting times:

Γ				i	1	2	3
0.000	0.000	1.000		δ_i	0.342	0.259	0.399
0.298	0.575	0.127		μ_i	55.30	75.30	84.93
0.662	0.276	0.062		σ_i	5.809	3.808	5.433

for $i = 1$ to 3. The marginal densities of the HMMs for the durations
are presented in Figure 10.2, and for the waiting times in Figure 10.3.

One feature of these models that is noticeable is that, although the
matrices $\mathbf{\Gamma}$ are by no means identical, in both cases the first row is
$(0,0,1)$ and the largest element of the third row is γ_{31}, the probability of
transition from state 3 to state 1.

10.4 Bivariate normal–HMM for durations and waiting times

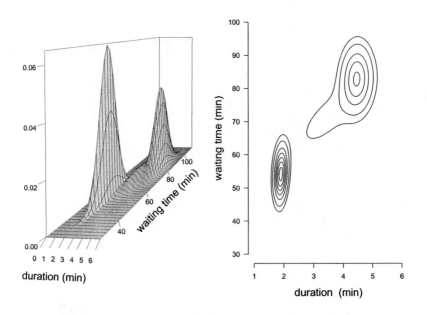

Figure 10.4 *Old Faithful durations and waiting times: perspective and contour plots of the p.d.f. of the bivariate normal–HMM.(Model fitted by discrete
likelihood.)*

Finally we give here a stationary three-state bivariate model for durations d_t and waiting times w_{t+1}, for t running from 1 to 298. The pairing
(d_t, w_t) would be another possibility; in Exercise 2 the reader is invited
to fit a similar model to that bivariate series and compare the two models. The three state-dependent distributions are general bivariate normal
distributions. Hence there are in all $21 = 15 + 6$ parameters: 5 for each
bivariate normal distribution, and 6 for the transition probabilities. The
model was fitted by maximizing the discrete likelihood of the bivariate

observations (d_t, w_{t+1}); that is, by maximizing

$$L_T = \delta\mathbf{P}(d_1, w_2)\mathbf{\Gamma P}(d_2, w_3)\cdots\mathbf{\Gamma P}(d_{T-1}, w_T)\mathbf{1}',$$

where $\mathbf{P}(d_t, w_{t+1})$ is the diagonal matrix with each diagonal element not a bivariate normal density but a probability: the probability that the tth pair (duration, waiting time) falls in the rectangle $(d_t^-, d_t^+) \times (w_{t+1}^-, w_{t+1}^+)$. Here d_t^- and d_t^+ represent the lower and upper bounds available for the tth duration, and similarly for waiting times. The code used can be found in A.3. Figure 10.4 displays perspective and contour plots of the marginal p.d.f. of this model, the parameters of which are given in Table 10.11. Note that here $\mathbf{\Gamma}$, the t.p.m., is of the same form as the matrices displayed in Table 10.10.

Table 10.11 *Old Faithful durations and waiting times: three-state bivariate normal–HMM, based on discrete likelihood.*

$\mathbf{\Gamma}$			i (state)	1	2	3
0.000	0.000	1.000	δ_i	0.283	0.229	0.488
0.037	0.241	0.722	mean duration	1.898	3.507	4.460
0.564	0.356	0.080	mean waiting time	54.10	71.59	83.18
			s.d. duration	0.142	0.916	0.322
			s.d. waiting time	4.999	8.289	6.092
			correlation	0.178	0.721	0.044

Exercises

1.(a) Use the code in A.3 to fit a bivariate normal–HMM with *two* states to the observations (d_t, w_{t+1}).

 (b) Compare the resulting marginal distributions for durations and waiting times to those implied by the three-state model reported in Table 10.11.

2. Fit a bivariate normal–HMM with three states to the observations (d_t, w_t), where t runs from 1 to 299. How much does this model differ from that reported in Table 10.11?

3.(a) Write an **R** function to generate observations from a bivariate normal–HMM. (Hint: see the code in A.2.1, and use the package mvtnorm.)

 (b) Write a function that will find MLEs of the parameters of a bivariate normal–HMM when the observations are assumed to be known exactly; use the 'continuous likelihood', i.e. use densities, not probabilities, in the likelihood.

(c) Use the function in 3(a) to generate a series of 1000 observations, and use the function in 3(b) to estimate the parameters.

(d) Apply varying degrees of interval censoring to your generated series and use the code in A.3 to estimate parameters. To what extent are the parameter estimates affected by interval censoring?

CHAPTER 11

Speed and change of direction in larvae of *Drosophila melanogaster*

11.1 Introduction

Holzmann *et al.* (2006) have described, *inter alia*, the application of HMMs with circular state-dependent distributions to the movement of larvae of the fruit fly *Drosophila melanogaster*. It is thought that locomotion can be largely summarized by the distribution of speed and direction change in each of two episodic states: 'forward peristalsis' (linear movement) and 'head swinging and turning' (Suster *et al.*, 2003). During linear movement, larvae maintain a high speed and a low direction change, in contrast to the low speed and high direction change characteristic of turning episodes. Given that the larvae apparently alternate thus between two states, an HMM in which both speed and turning rate are modelled according to two underlying states might be appropriate for describing the pattern of larval locomotion. As illustration we shall examine the movements of two of the experimental subjects of Suster (2000) (one wild larva, one mutant) whose positions were recorded once per second. The paths taken by the larvae are displayed in Figure 11.1.

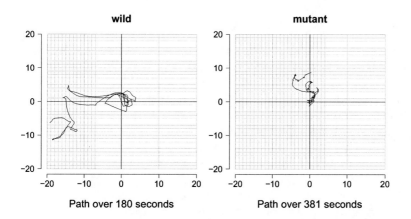

Figure 11.1 *Paths of two larvae of* Drosophila melanogaster.

First we examine the univariate time series of direction changes for the two larvae; then in Section 11.5 we shall examine the bivariate series of speeds and direction changes.

11.2 Von Mises distributions

We need to discuss here an important family of distributions designed for circular data, the von Mises distributions, which have properties that make them in some respects a natural first choice as a model for unimodal continuous observations on the circle; see e.g. Fisher (1993, pp. 49–50, 55). The probability density function of the von Mises distribution with parameters $\mu \in (-\pi, \pi]$ (location) and $\kappa > 0$ (concentration) is

$$f(x) = (2\pi I_0(\kappa))^{-1} \exp(\kappa \cos(x - \mu)) \quad \text{for } x \in (-\pi, \pi]. \tag{11.1}$$

Here I_0 denotes the modified Bessel function of the first kind of order zero. More generally, for integer n, I_n is given in integral form by

$$I_n(\kappa) = (2\pi)^{-1} \int_{-\pi}^{\pi} \exp(\kappa \cos x) \cos(nx) \, dx; \tag{11.2}$$

see e.g. Equation (9.6.19) of Abramowitz *et al.* (1984). But note that one could in the p.d.f. (11.1) replace the interval $(-\pi, \pi]$ by $[0, 2\pi)$, or by any other interval of length 2π.

The location parameter μ is not in the usual sense the mean of a random variable X having the above density; instead (see Exercise 3) it satisfies

$$\tan \mu = E(\sin X)/E(\cos X),$$

and can be described as a directional mean, or circular mean. We use the convention that $\arctan(b, a)$ is the angle $\theta \in (-\pi, \pi]$ such that $\tan \theta = b/a$, $\sin \theta$ and b have the same sign, and $\cos \theta$ and a have the same sign. (But note that if $b = a = 0$, $\arctan(b, a)$ is not defined.) With this convention, $\mu = \arctan(E \sin X, E \cos X)$, and indeed this is the definition we use for the directional mean of any circular random variable X, not only one having a von Mises distribution. The sample equivalent of μ, based on the sample $x_1, x_2, \ldots, x_T$, is $\hat{\mu} = \arctan(\sum_t \sin x_t, \sum_t \cos x_t)$.

In modelling time series of directional data one can consider using an HMM with von Mises distributions as the state-dependent distributions, although any other circular distribution is also possible. For several other classes of models for time series of directional data, based on ARMA processes, see Fisher and Lee (1994).

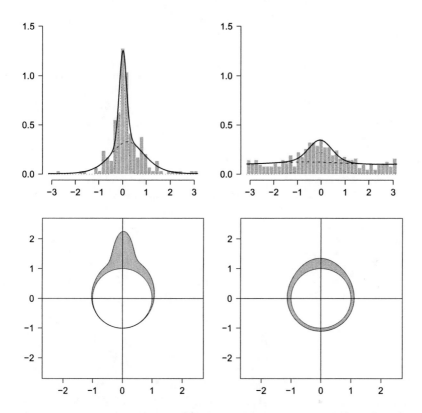

Figure 11.2 *Marginal density plots for two-state von Mises–HMMs for direction change in* Drosophila melanogaster: *wild subject (left) and mutant (right). In the plots in the first row angles are measured in radians, and the state-dependent densities, multiplied by their mixing probabilities* δ_i, *are indicated by dashed and dotted lines. In the circular density plots in the second row, North corresponds to a zero angle, right to positive angles, left to negative.*

11.3 Von Mises–HMMs for the two subjects

Here we present a two-state von Mises–HMM for the series of changes of direction for each of the subjects, and we compare those models, on the basis of likelihood, AIC and BIC, with the corresponding three-state models which we have also fitted. Table 11.1 presents the relevant comparison. For the wild subject, both AIC and BIC suggest that the two-state model is adequate, but they disagree in the case of the mutant; BIC selects the two-state and AIC the three-state.

The models are displayed in Table 11.2, with μ_i, for instance, denoting

Table 11.1 *Comparison of two- and three-state von Mises–HMMs for changes of direction in two* Drosophila melanogaster *larvae.*

subject	no. states	no. parameters	$-\log L$	AIC	BIC
wild	2	6	158.7714	**329.5**	**348.7**
	3	12	154.4771	333.0	371.3
mutant	2	6	647.6613	1307.3	**1331.0**
	3	12	638.6084	**1301.2**	1348.5

Table 11.2 *Two-state von Mises–HMMs for the changes of direction in two* Drosophila melanogaster *larvae.*

	wild, state 1	wild, state 2	mutant, state 1	mutant, state 2
μ_i	0.211	0.012	−0.613	−0.040
κ_i	2.050	40.704	0.099	4.220
Γ	$\begin{pmatrix} 0.785 & 0.215 \\ 0.368 & 0.632 \end{pmatrix}$		$\begin{pmatrix} 0.907 & 0.093 \\ 0.237 & 0.763 \end{pmatrix}$	
δ_i	0.632	0.368	0.717	0.283

the location parameter in state i, and the marginal densities are depicted in Figure 11.2. Clearly there are marked differences between the two subjects, for instance the fact that one of the states of the wild subject, that with mean close to zero, has a very high concentration.

11.4 Circular autocorrelation functions

As a diagnostic check of the fitted two-state models one can compare the autocorrelation functions of the model with the sample autocorrelations. In doing so, however, one must take into account the circular nature of the observations.

There are (at least) two proposed measures of correlation of circular observations: one due to Fisher and Lee (1983), and another due to Jammalamadaka and Sarma (1988), which we concentrate on. Using the latter one can find the ACF of a fitted von Mises–HMM and compare it with its empirical equivalent, and we have done so for the direction change series of the two subjects. It turns out that there is little or no serial correlation in these series of angles, but there is non-negligible ordinary autocorrelation in the series of absolute values of the direction changes.

The measure of circular correlation proposed by Jammalamadaka and Sarma is as follows (see Jammalamadaka and SenGupta, 2001, (8.2.2), p. 176). For two (circular) random variables Θ and Φ, the circular correlation is defined as

$$\frac{E(\sin(\Theta - \mu_\theta)\sin(\Phi - \mu_\phi))}{[\text{Var}\sin(\Theta - \mu_\theta)\,\text{Var}\sin(\Phi - \mu_\phi)]^{1/2}},$$

where μ_θ is the directional mean of Θ, and similarly μ_ϕ that of Φ. Equivalently, it is

$$\frac{E(\sin(\Theta - \mu_\theta)\sin(\Phi - \mu_\phi))}{[E\sin^2(\Theta - \mu_\theta)\,E\sin^2(\Phi - \mu_\phi)]^{1/2}};$$

see Exercise 1 for justification.

The autocorrelation function of a (stationary) directional series X_t is then defined by

$$\rho(k) = \frac{E(\sin(X_t - \mu)\sin(X_{t+k} - \mu))}{[E\sin^2(X_t - \mu)\,E\sin^2(X_{t+k} - \mu)]^{1/2}};$$

this simplifies to

$$\rho(k) = \frac{E(\sin(X_t - \mu)\sin(X_{t+k} - \mu))}{E\sin^2(X_t - \mu)}. \tag{11.3}$$

Here μ denotes the directional mean of X_t (and X_{t+k}), that is

$$\mu = \arctan(E\sin X_t, E\cos X_t).$$

An estimator of the ACF is given by

$$\sum_{t=1}^{T-k}\sin(x_t - \hat{\mu})\sin(x_{t+k} - \hat{\mu}) \Big/ \sum_{t=1}^{T}\sin^2(x_t - \hat{\mu}),$$

with $\hat{\mu}$ denoting the sample directional mean of all T observations x_t.

With some work it is also possible to to use Equation (11.3) to compute the ACF of a von Mises–HMM as follows. Firstly, note that, by Equation (2.10), the numerator of (11.3) is

$$\sum_{i=1}^{m}\sum_{j=1}^{m}\delta_i\gamma_{ij}(k)E(\sin(X_t - \mu)\mid C_t = i)E(\sin(X_{t+k} - \mu)\mid C_{t+k} = j),$$

$$\tag{11.4}$$

and that, if $C_t = i$, the observation X_t has a von Mises distribution with parameters μ_i and κ_i. To find the conditional expectation of $\sin(X_t - \mu)$ given $C_t = i$ it is therefore convenient to write

$$
\begin{aligned}
\sin(X_t - \mu) &= \sin(X_t - \mu_i + \mu_i - \mu) \\
&= \sin(X_t - \mu_i)\cos(\mu_i - \mu) + \cos(X_t - \mu_i)\sin(\mu_i - \mu).
\end{aligned}
$$

The conditional expectation of $\cos(X_t - \mu_i)$ is $I_1(\kappa_i)/I_0(\kappa_i)$, and that of $\sin(X_t - \mu_i)$ is zero; see Exercise 2. The conditional expectation of $\sin(X_t - \mu)$ given $C_t = i$ is therefore available, and similarly that of $\sin(X_{t+k} - \mu)$ given $C_{t+k} = j$. Hence expression (11.4), i.e. the numerator of (11.3), can be computed once one has μ.

Secondly, note that, by Exercise 4, the denominator of (11.3) is

$$\mathrm{E}\sin^2(X_t - \mu) = \frac{1}{2}\left(1 - \sum_i \delta_i A_2(\kappa_i) \cos(2(\mu_i - \mu))\right),$$

where, for positive integers n, $A_n(\kappa)$ is defined by $A_n(\kappa) = I_n(\kappa)/I_0(\kappa)$.

All that now remains is to indicate how to compute μ for such a von Mises–HMM. This is given by

$$\mu = \arctan\left(\mathrm{E}\sin X_t, \mathrm{E}\cos X_t\right) = \arctan\left(\sum_i \delta_i \sin\mu_i, \sum_i \delta_i \cos\mu_i\right).$$

The ACF, $\rho(k)$, can therefore be found as follows:

$$\rho(k) = \frac{\sum_i \sum_j \delta_i \gamma_{ij}(k) A_1(\kappa_i) \sin(\mu_i - \mu) A_1(\kappa_j) \sin(\mu_j - \mu)}{\frac{1}{2}(1 - \sum_i \delta_i A_2(\kappa_i) \cos(2(\mu_i - \mu)))}.$$

As a function of k, this is of the usual form taken by the ACF of an HMM; provided the eigenvalues of $\mathbf{\Gamma}$ are distinct, it is a linear combination of the kth powers of these eigenvalues.

Earlier Fisher and Lee (1983) proposed a slightly different correlation coefficient between two circular random variables. This definition can also be used to provide a corresponding (theoretical) ACF for a circular time series, for which an estimator is provided by Fisher and Lee (1994), Equation (3.1). For details, see Exercises 5 and 6.

We have computed the Jammalamadaka–Sarma circular ACFs of the two series of direction changes, and found little or no autocorrelation. Furthermore, for series of the length we are considering here, and circular autocorrelations (by whichever definition) that are small or very small, the two estimators of circular autocorrelation appear to be highly variable.

An autocorrelation that does appear to be informative, however, is the (ordinary) ACF of the *absolute values* of the series of direction changes; see Figure 11.3. Although there is little or no circular autocorrelation in each series of direction changes, there is non-negligible ordinary autocorrelation in the series of absolute values of the direction changes, especially in the case of the mutant. This is rather similar to the 'stylized fact' of share return series that the ACF of returns is negligible, but not the ACF of absolute or squared returns. We can therefore not conclude that the direction changes are (serially) independent.

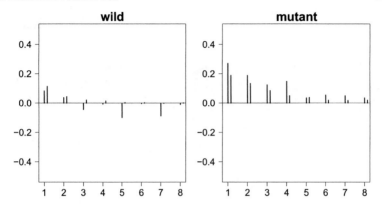

Figure 11.3 *ACFs of absolute values of direction changes. In each pair of ACF values displayed, the left one is the (ordinary) autocorrelation of the absolute changes of direction of the* sample, *and the right one is the corresponding quantity based on 50 000 observations generated from the two-state* model.

11.5 A bivariate series with one component linear and one circular: *Drosophila* speed and change of direction

We begin our analysis of the bivariate time series of speed and change of direction (c.o.d.) for *Drosophila* by examining a scatter plot of these quantities for each of the two subjects (top half of Figure 11.4). A smooth of these points is roughly horizontal, but the funnel shape of the plot in each case is conspicuous. We therefore plot also the speeds and absolute changes of direction, and we find, as one might expect from the funnel shape, that a smooth now has a clear downward slope.

We have also plotted in each of these figures a smooth of 10 000 points generated from the three-state model we describe later in this section. In only one of the four plots do the two lines differ appreciably, that of speed and absolute c.o.d. for the wild subject (lower left); there the line based on the model is the higher one. We defer further comment on the models to later in this section.

The structure of the HMMs fitted to this bivariate series is as follows. Conditional on the underlying state, the speed at time t and the change of direction at time t are assumed to be independent, with the former having a gamma distribution and the latter a von Mises distribution. In the three-state model there are in all 18 parameters to be estimated: six transition probabilities, two parameters for each of the three gamma distributions, and two parameters for each of the three von Mises distributions; more generally, $m^2 + 3m$ for an m-state model of this kind.

These and other models for the *Drosophila* data were fairly difficult

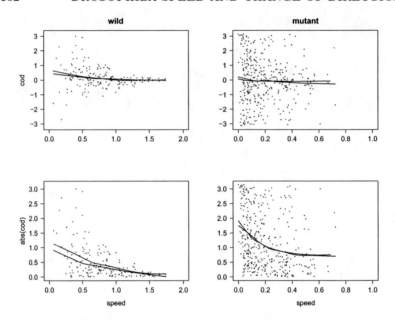

Figure 11.4 *Each panel shows the observations (change of direction against speed in the top panels, and absolute change of direction against speed in the bottom panels). In each case there are two nonparametric regression lines computed by the* **R** *function* **loess**, *one smoothing the observations and the other smoothing the 10 000 realizations generated from the fitted model.*

Table 11.3 *Comparison of two- and three-state bivariate HMMs for speed and direction change in two larvae of* Drosophila.

subject	no. states	no. parameters	$-\log L$	AIC	BIC
wild	2	10	193.1711	406.3	438.3
	3	18	166.8110	**369.6**	**427.1**
mutant	2	10	332.3693	684.7	724.2
	3	18	303.7659	**643.5**	**714.5**

to fit in that it was easy to become trapped at a local optimum of the log-likelihood which was not the global optimum.

Table 11.3 compares the two- and three-state models on the basis of likelihood, AIC and BIC, and indicates that AIC and BIC select three states, both for the wild subject and for the mutant. Figure 11.5 depicts the marginal and state-dependent distributions of the three-state

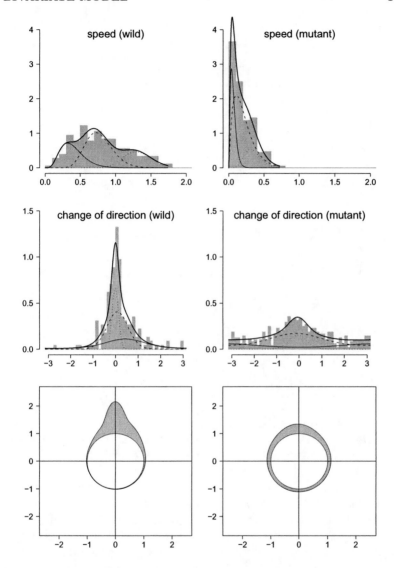

Figure 11.5 *Three-state (gamma–von Mises) HMMs for speed and change of direction in* Drosophila melanogaster: *wild subject (left) and mutant (right). The top panels show the marginal and the state-dependent (gamma) distributions for the speed, as well as a histogram of the speeds. The middle panels show the marginal and the state-dependent (von Mises) distributions for the c.o.d., and a histogram of the changes of direction. The bottom panels show the fitted marginal densities for c.o.d.*

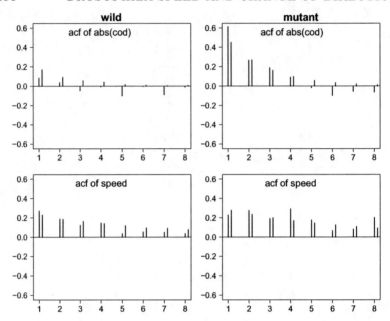

Figure 11.6 *Sample and 3-state-HMM ACFs of absolute change of direction (top panels) and speed (bottom panels). The ACFs for the absolute change of direction under the model were computed by simulation of a series of length 50 000.*

model. Figure 11.6 compares the sample and three-state model ACFs for absolute c.o.d. and for speed; for the absolute c.o.d. the model ACF was estimated by simulation, and for the speed it was computed by using the results of Exercise 3 of Chapter 2.

What is noticeable about the models for c.o.d. depicted in Figure 11.5 is that, although not identical, they are visually very similar to the models of Figure 11.2, even though their origin is rather different. In one case there are two states, in the other three; in one case the model is univariate, and in the other it is one of the components of a bivariate model.

This application illustrates two attractive features of HMMs as models for bivariate or multivariate time series. The first is the ease with which different data types can be accommodated — here we modelled a bivariate series with one circular-valued and one (continuous) linear-valued variable. Similarly it is possible to model bivariate series in which one variable is discrete-valued (even categorical) and the other continuous-valued. The second feature is that the assumption of contemporaneous conditional independence provides a very simple means of modelling dependence between the contemporaneous values of the component series, whether these are bivariate or multivariate. Although the range of dependence structures which can be accommodated in this way is limited, it seems to be adequate in some applications.

Exercises

1. Let μ be the directional mean of the circular random variable Θ, as defined on p. 156. Show that $E \sin(\Theta - \mu) = 0$.

2. Let X have a von Mises distribution with parameters μ and κ. Show that

$$E \cos(n(X - \mu)) = I_n(\kappa)/I_0(\kappa) \text{ and } E \sin(n(X - \mu)) = 0,$$

and hence that

$$E \cos(nX) = \cos(n\mu) I_n(\kappa)/I_0(\kappa)$$

and

$$E \sin(nX) = \sin(n\mu) I_n(\kappa)/I_0(\kappa).$$

More compactly,

$$E(e^{inX}) = e^{in\mu} I_n(\kappa)/I_0(\kappa).$$

3. Again, let X have a von Mises distribution with parameters μ and κ. Deduce from the conclusions of Exercise 2 that

$$\tan \mu = E(\sin X)/E(\cos X).$$

4. Let $\{X_t\}$ be a stationary von Mises–HMM on m states, with the ith state-dependent distribution being von Mises (μ_i, κ_i), and with μ denoting the directional mean of X_t.

   Show that

$$E \sin^2(X_t - \mu) = \frac{1}{2}\left(1 - \sum_i \delta_i A_2(\kappa_i) \cos(2(\mu_i - \mu))\right),$$

where $A_2(\kappa) = I_2(\kappa)/I_0(\kappa)$. Hint: use the following steps.

$$E \sin^2(X_t - \mu) = \sum \delta_i E(\sin^2(X_t - \mu) \mid C_t = i);$$

$$\sin^2 A = (1 - \cos(2A))/2; \quad X_t - \mu = X_t - \mu_i + \mu_i - \mu;$$

$$E(\sin(2(X_t - \mu_i)) \mid C_t = i) = 0;$$

and

$$E(\cos(2(X_t - \mu_i)) \mid C_t = i) = I_2(\kappa_i)/I_0(\kappa_i).$$

Is it necessary to assume that μ is the directional mean of X_t?

5. Consider the definition of the circular correlation coefficient given by Fisher and Lee (1983): the correlation between two circular random variables Θ and Φ is

$$\frac{E(\sin(\Theta_1 - \Theta_2)\sin(\Phi_1 - \Phi_2))}{[E(\sin^2(\Theta_1 - \Theta_2))E(\sin^2(\Phi_1 - \Phi_2))]^{1/2}},$$

where (Θ_1, Φ_1) and (Θ_2, Φ_2) are two independent realizations of (Θ, Φ).

Show that this definition implies the following expression, given as Equation (2) of Holzmann *et al.* (2006), for the circular autocorrelation of order k of a stationary time series of circular observations X_t:

$$\frac{E(\cos X_0 \cos X_k)E(\sin X_0 \sin X_k) - E(\sin X_0 \cos X_k)E(\cos X_0 \sin X_k)}{(1 - E(\cos^2 X_0))E(\cos^2 X_0) - (E(\sin X_0 \cos X_0))^2}.$$

$$(11.5)$$

6. Show how Equation (11.5), along with Equations (2.8) and (2.10), can be used to compute the Fisher–Lee circular ACF of a von Mises–HMM.

CHAPTER 12

Wind direction at Koeberg

12.1 Introduction

South Africa's only nuclear power station is situated at Koeberg on the west coast, about 30 km north of Cape Town. Wind direction, wind speed, rainfall and other meteorological data are collected continuously by the Koeberg weather station with a view to their use in radioactive plume modelling, *inter alia*. Four years of data were made available by the staff of the Koeberg weather station, and this chapter describes an attempt to model the wind direction at Koeberg by means of HMMs.

The wind direction data consist of hourly values of average wind direction over the preceding hour at 35 m above ground level. The period covered is 1 May 1985 to 30 April 1989 inclusive. The average referred to is a vector average, which allows for the circular nature of the data, and is given in degrees. There are in all 35 064 observations; there are no missing values.

12.2 Wind direction classified into 16 categories

Although the hourly averages of wind direction were available in degrees, the first group of models fitted treated the observations as lying in one of the 16 conventional directions N, NNE, ..., NNW, coded 1 to 16 in that order. This was done in order to illustrate the application of HMMs to time series of categorical observations, a special class of multinomial-like time series.

12.2.1 Three HMMs for hourly averages of wind direction

The first model fitted was a simple multinomial–HMM with two states and no seasonal components, the case $m = 2$ and $q = 16$ of the categorical model described in Section 8.4.2. In this model there are 32 parameters to be estimated: two transition probabilities to specify the Markov chain, and 15 probabilities for each of the two states, subject to the sum of the 15 not exceeding one. The results are as follows. The underlying Markov chain has transition probability matrix $\begin{pmatrix} 0.964 & 0.036 \\ 0.031 & 0.969 \end{pmatrix}$ and stationary distribution (0.462, 0.538), and the 16 probabilities associated with each

Table 12.1 *Koeberg wind data (hourly): 1000×probabilities of each direction in each state for (from left to right) the simple two-state HMM, the simple three-state HMM, and the two-state HMM with cyclical components.*

		2-state HMM		3-state HMM			cyclic HMM	
1	N	129	0	148	0	1	127	0
2	NNE	48	0	47	0	16	47	0
3	NE	59	1	16	0	97	57	2
4	ENE	44	26	3	0	148	27	40
5	E	6	50	1	0	132	4	52
6	ESE	1	75	0	0	182	1	76
7	SE	0	177	0	23	388	1	179
8	SSE	0	313	0	426	33	0	317
9	S	1	181	0	257	2	1	183
10	SSW	4	122	2	176	0	7	121
11	SW	34	48	20	89	0	59	26
12	WSW	110	8	111	28	0	114	3
13	W	147	0	169	2	0	145	0
14	WNW	130	0	151	0	0	128	0
15	NW	137	0	159	0	1	135	0
16	NNW	149	0	173	0	0	147	0

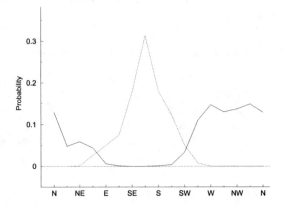

Figure 12.1 *Koeberg wind data (hourly): probabilities of each direction in the simple two-state HMM.*

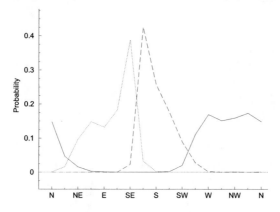

Figure 12.2 *Koeberg wind data (hourly): probabilities of each direction in the three-state HMM.*

of the two states are displayed in columns 3 and 4 of Table 12.1. A graph of these two sets of probabilities appears as Figure 12.1. The value of the unconditional log-likelihood achieved by this model is $-75\,832.1$.

The model successfully identifies two apparently meaningful weather states which are very different at least as regards the likely wind directions in those states. In state 1 the most likely direction is NNW, and the probability falls away on either side of NNW, reaching a level of less than 0.01 for all directions (clockwise) from E to SSW inclusive. In state 2 the peak is at direction SSE, and the probability falls away sharply on either side, being less than 0.01 for all directions from WSW to NE inclusive. Two generalizations of this type of model were also fitted: firstly, a model based on a three-state Markov chain rather than a two-state one; and secondly, a model based on a two-state Markov chain but incorporating both a daily cycle and an annual cycle. We shall now describe these two models, and in due course compare all the models considered on the basis of AIC and BIC.

The three-state HMM has 51 parameters: six to specify the Markov chain, and 15 probabilities associated with each of the states. Essentially this model splits state 2 of the two-state model into two new states, one of which peaks at SSE and the other at SE. The transition probability matrix is

$$\begin{pmatrix} 0.957 & 0.030 & 0.013 \\ 0.015 & 0.923 & 0.062 \\ 0.051 & 0.077 & 0.872 \end{pmatrix}$$

and the stationary distribution is $(0.400, 0.377, 0.223)$. The 16 probabilities associated with each of the three states are displayed in Figure 12.2 and columns 5–7 of Table 12.1. The unconditional log-likelihood is $-69\,525.9$.

As regards the model which adds daily and annual cyclical effects to the simple two-state HMM, it was decided to build these effects into the Markov chain rather than into the state-dependent probabilities. This was because, for a two-state chain, we can model cyclical effects parsimoniously by assuming that the two off-diagonal transition probabilities

$$\Pr(C_t \neq i \mid C_{t-1} = i) = {}_t\gamma_{i,3-i} \quad (i = 1, 2)$$

are given by an appropriate periodic function of t. We assume that logit ${}_t\gamma_{i,3-i}$ is equal to

$$a_i + \underbrace{b_i \cos(2\pi t/24) + c_i \sin(2\pi t/24)}_{\text{daily cycle}} + \underbrace{d_i \cos(2\pi t/8766) + e_i \sin(2\pi t/8766)}_{\text{annual cycle}}$$

$$(12.1)$$

for $i = 1, 2$ and $t = 2, 3, \ldots, T$. A similar model for each of the state-dependent probabilities in each of the two states would involve many more parameters. As discussed in Section 8.5.2, the estimation technique has to be modified when the underlying Markov chain is not assumed to be homogeneous. The estimates in this case were based on the initial state of the Markov chain being state 2, and the log of the likelihood conditioned on that initial state is $-75\,658.5$. (Conditioning on state 1 yielded a slightly inferior value for the likelihood, and similar parameter estimates.)

The probabilities associated with each state are given in the last two columns of Table 12.1, and the estimated parameters for the two off-diagonal transition probabilities are given below, in the notation of Equation (12.1).

i	a_i	b_i	c_i	d_i	e_i
1	−3.349	0.197	−0.695	−0.208	−0.401
2	−3.523	−0.272	0.801	0.082	−0.089

From Table 12.1 it will be noted that the general pattern of the state-dependent probabilities is in this case very similar to that of the simple two-state model without any cyclical components.

12.2.2 Model comparisons and other possible models

The three models described above were compared with each other and with a saturated 16-state Markov chain model, on the basis of AIC and BIC. The transition probabilities defining the Markov chain model were estimated by conditional maximum likelihood, as described in Section

Table 12.2 *Koeberg wind data (hourly):* $1000\times$ *transition probability matrix of saturated Markov chain model.*

610	80	22	8	3	1	1	2	4	0	3	3	14	20	37	190
241	346	163	37	15	1	3	3	3	9	10	29	37	36	64
56	164	468	134	28	6	9	6	4	6	6	18	32	18	19	25
13	33	163	493	144	32	17	11	11	17	7	12	14	14	10	10
9	7	48	249	363	138	53	27	33	22	11	6	8	7	13	9
4	9	10	51	191	423	178	51	25	23	8	7	4	6	4	4
1	1	3	6	23	141	607	160	36	8	6	3	2	1	1	1
1	1	1	3	5	16	140	717	94	13	5	3	2	1	0	0
4	1	2	4	5	8	25	257	579	77	17	9	5	3	1	0
2	2	2	1	6	5	10	36	239	548	93	41	10	5	2	0
5	2	3	5	3	3	8	12	38	309	397	151	38	12	8	5
4	2	2	3	1	3	2	5	17	56	211	504	149	19	16	7
10	5	5	3	4	1	2	5	4	13	28	178	561	138	30	13
13	5	4	3	3	3	1	1	1	7	8	27	188	494	199	43
31	9	7	4	5	2	1	1	2	1	3	11	43	181	509	190
158	23	9	5	1	1	2	1	0	2	2	4	17	54	162	559

Table 12.3 *Koeberg wind data (hourly): comparison of four models fitted.*

model	k	$-l$	AIC	BIC
2-state HMM	32	75 832.1	151 728	151 999
3-state HMM	51	69 525.9	139 154	139 585
2-state HMM with cycles	40	75 658.5*	151 397	151 736
saturated Markov chain	240	48 301.7	**97 083**	**99 115**

* conditional on state 2 being the initial state

1.3.5, and are displayed in Table 12.2. The comparison appears as Table 12.3.

What is of course striking in Table 12.3 is that the likelihood of the saturated Markov chain model is so much higher than that of the HMMs that the large number of parameters of the Markov chain (240) is virtually irrelevant when comparisons are made by AIC or BIC. It is therefore interesting to compare certain properties of the Markov chain model with the corresponding properties of the simple two-state HMM (e.g. unconditional and conditional probabilities). The unconditional probabilities of each direction were computed for the two models, and are almost identical.

However, examination of $\Pr(X_{t+1} = 16 \mid X_t = 8)$, where X_t denotes the direction at time t, points to an important difference between the

Table 12.4 *Koeberg wind data (daily at hour 1): comparison of two models fitted.*

model	k	$-l$	AIC	BIC
2-state HM	32	3461.88	**6987.75**	**7156.93**
saturated MC	240	3292.52	7065.04	8333.89

models. For the Markov chain model this probability is zero, since no such transitions were observed. For the HMM it is, in the notation of Equation (8.1),

$$\frac{\boldsymbol{\delta}\boldsymbol{\Pi}(8)\boldsymbol{\Gamma}\boldsymbol{\Pi}(16)\mathbf{1}'}{\boldsymbol{\delta}\boldsymbol{\Pi}(8)\mathbf{1}'} = \frac{0.5375 \times 0.3131 \times 0.0310 \times 0.1494}{0.5375 \times 0.3131} = 0.0046.$$

Although small, this probability is not insignificant. The observed number of transitions from SSE (direction 8) was 5899. On the basis of the HMM one would therefore expect about 27 of these transitions to be to NNW. None were observed. Under the HMM 180-degree switches in direction are quite possible. Every time the state changes (which happens at any given time point with probability in excess of 0.03), the most likely direction of wind changes by 180 degrees. This is inconsistent with the observed gradual changes in direction; the matrix of observed transition counts is heavily dominated by diagonal and near-diagonal elements (and, because of the circular nature of the categories, elements in the corners furthest from the principal diagonal). Changes through 180 degrees were in general rarely observed.

The above discussion suggests that, if daily figures are examined rather than hourly, the observed process will be more amenable to modelling by means of an HMM, because abrupt changes of direction are more likely in daily data than hourly. A Markov chain and a two-state HMM were therefore fitted to the series of length 1461 beginning with the first observation and including every 24th observation thereafter. For these data the HMM proved to be superior to the Markov chain even on the basis of AIC, which penalizes extra parameters less here than does BIC: see Table 12.4. Although a daily model is of little use in the main application intended (evacuation planning), there are other applications for which a daily model is exactly what one needs, e.g. forecasting the wind direction a day ahead.

Since the (first-order) Markov chain model does not allow for dependence beyond first order, the question that arises is whether any model for the hourly data which allows for higher-order dependence is superior to the Markov chain model; the HMMs considered clearly are not.

Table 12.5 *Koeberg wind data (hourly): comparison of first-order Markov chain with lag-2 Raftery models.*

model	k	$-l$	AIC	BIC
saturated MC	240	48 301.7	97 083.4	99 115.0
Raftery model with starting values	241	48 087.8*	96 657.5	98 697.6
Raftery model fitted by max. likelihood	241	48 049.8*	**96 581.6**	**98 621.7**

* conditioned on the first two states

A saturated second-order Markov chain model would have an excessive number of parameters, and the Pegram model for (e.g.) a second-order Markov chain cannot reflect the property that, if a transition is made out of a given category, a nearby category is a more likely destination than a distant one.

The Raftery models (also known as MTD models: see p. 23) do not suffer from that disadvantage, and in fact it is very easy to find a lag-2 Raftery model that is convincingly superior to the first-order Markov chain model. In the notation of Section 1.3.6, take $\mathbf{Q}$ to be the transition probability matrix of the first-order Markov chain, as displayed in Table 12.2, and perform a simple line-search to find that value of λ_1 which maximizes the resulting conditional likelihood. This turns out to be 0.925. With these values for $\mathbf{Q}$ and λ_1 as starting values, the conditional likelihood was then maximized with respect to all 241 parameters, subject to the assumption that $0 \leq \lambda_1 \leq 1$. (This assumption makes it unnecessary to impose 16^3 pairs of nonlinear constraints on the maximization, and seems reasonable in the context of hourly wind directions showing a high degree of persistence.) The resulting value for λ_1 is 0.9125, and the resulting matrix $\mathbf{Q}$ does not differ much from its starting value. The log-likelihood achieved is $-48\,049.8$: see Table 12.5 for a comparison of likelihood values and the usual model selection criteria, from which the Raftery model emerges as superior to the Markov chain. Berchtold (2001) reports that this model cannot be improved on by another MTD model.

12.2.3 Conclusion

The conclusion we may draw from the above analysis is that, both for hourly and daily (categorized) wind direction data, it is possible to fit a model which is superior (according to the model selection criteria used) to a saturated first-order Markov chain. In the case of the hourly data, a lag-2 Raftery model (i.e. a particular kind of second-order Markov chain model) is preferred. In the case of the daily data, a simple two-state HMM for categorical time series, as introduced in Section 8.4.2, performs better than the Markov chain.

12.3 Wind direction as a circular variable

We now revisit the Koeberg wind direction data, with a view to demonstrating some of the many variations of HMMs that could be applied to data such as these. In particular, we do not confine ourselves here to use of the directions as classified into the 16 points of the compass, and we make use of the corresponding observations of wind speed in several different ways as a covariate in models for the direction, or change in direction. Some of the models are not particularly successful, but it is not our intention here to confine our attention to models which are in some sense the best, but rather to demonstrate the versatility of HMMs and to explore some of the questions which may arise in the fitting of more complex models.

12.3.1 Daily at hour 24: von Mises–HMMs

The analysis of wind direction that we have so far described is based entirely on the data as classified into the 16 directions N, NNE, ..., NNW. Although it illustrates the fitting of models to categorical series, it does ignore the fact that the observations of (hourly average) direction are available in degrees, and it ignores the circular nature of the observations.

One of the strengths of the hidden Markov formulation is that almost any kind of data can be accommodated at observation level. Here we can exploit that flexibility by taking the state-dependent distribution from a family of distributions designed for circular data, the von Mises distributions. The probability density function of the von Mises distribution with parameters μ and κ is

$$f(x) = (2\pi I_0(\kappa))^{-1} \exp(\kappa \cos(x - \mu)) \quad \text{for } x \in (-\pi, \pi];$$

see also Section 11.2.

In this section we shall fit to the directions data (in degrees) HMMs which have 1–4 states and von Mises state-dependent distributions. This is perhaps a fairly obvious approach if one has available 'continuous' circular data, e.g. in integer degrees. We do this for the series of length 1461 formed by taking every 24th observation of direction. By taking

Table 12.6 *Koeberg wind data (daily at hour 24): comparison of models fitted to wind direction. m is the number of states in an HMM, k is in general the number of parameters in a model, and equals $m^2 + m$ for all but the Markov chain model. Note that the figures in the $-l$ column for the continuous models have to be adjusted before they can be compared with those for the discretized models.*

model	m	k	$-l$	$-l$(adjusted)	AIC	BIC
von Mises–HM	1	2	2581.312	3946.926	7897.9	7908.4
von Mises–HM	2	6	2143.914	3509.528	7031.1	7062.8
von Mises–HM	3	12	2087.133	3452.747	6929.5	6992.9
von Mises–HM	4	20	2034.133	3399.747	**6839.5**	**6945.2**
discretized von M–HM	1	2	3947.152		7898.3	7908.9
discretized von M–HM	2	6	3522.848		7057.7	7089.4
discretized von M–HM	3	12	3464.491		6953.0	7016.4
discretized von M–HM	4	20	3425.703		6891.4	6997.1
saturated Markov chain		240	3236.5		6953.0	8221.9

every 24th observation we are focusing on the hour each day from 23:00 to 24:00, and modelling the average direction over that hour.

It is somewhat less obvious that it is still possible to fit von Mises–HMMs if one has available only the categorized directions; that is a second family of models which we present here. To fit such models, one uses in the likelihood computation not the von Mises density but the integral thereof over the interval corresponding to the category observed. For the purpose of comparison we present also a 16-state saturated Markov chain fitted to the same series.

A minor complication that arises if one wishes to compare the continuous log-likelihoods with the discrete is that one has to bear in mind that the direct use of a density in a continuous likelihood is just a convention. The density should in fact be integrated over the smallest interval that can contain the observation, or the integral approximated by the density multiplied by the appropriate interval length. The effect here is that, if one wishes to compare the continuous log-likelihoods with those relating to to the 16 categories, one has to add $1461 \log(2\pi/16) = -1365.614$ to the continuous log-likelihood, or equivalently subtract it from minus the log-likelihood l. The resulting adjusted values of $-l$ appear in the column headed '$-l$(adjusted)' in Table 12.6.

12.3.2 Modelling hourly change of direction

Given the strong persistence of wind direction which is apparent from
the transition probability matrix in Table 12.2, the most promising approach, however, seems to be to model the *change* in direction rather
then the direction itself. We therefore describe here a variety of models for the change in (hourly average) direction from one hour to the
next, both with and without the use of wind speed, lagged one hour,
as a covariate. Observations of wind speed, in cm s^{-1}, were available
for the same period as the observations of direction. The use of lagged
wind speed as a covariate, rather than simultaneous, is motivated by
the need for any covariate to be available at the time of forecast; here a
1-hour-ahead forecast is of interest.

12.3.3 Transition probabilities varying with lagged speed

Change of direction, like direction itself, is a circular variable, and first
we model it by means of a von Mises–HMM with two, three, or four
states. Then we introduce wind speed, lagged one hour, into the model
as follows.

The usual transformation of the transition probabilities in row i (say)
is

$$\gamma_{ij} = e^{\tau_{ij}} \big/ \left(1 + \textstyle\sum_{k \neq i} e^{\tau_{ik}}\right) \quad j \neq i,$$

with γ_{ii} then determined by the row-sum constraint. Here we use instead
the transformation

$$\gamma_{ij} = \Pr(C_t = j \mid C_{t-1} = i, S_{t-1} = s) = e^{\tau_{ij}} \big/ \textstyle\sum_{k=1}^{m} e^{\tau_{ik}} \quad j = 1, 2, \ldots m$$

where

$$\tau_{ii} = \eta_i s$$

and S_{t-1} is the speed at time $t - 1$. Equivalently, the transformation is,
for row i,

$$\gamma_{ij} = e^{\tau_{ij}} \big/ \left(e^{\eta_i s} + \textstyle\sum_{k \neq i} e^{\tau_{ik}}\right) \quad j \neq i,$$

with γ_{ii} determined by the row-sum constraint. This structure allows
the speed at time $t - 1$ to influence the probabilities of transition from
state i to j between times $t - 1$ and t.

In passing, we note that there would be no point in introducing an
intercept η_{0i} as follows:

$$\tau_{ii} = \eta_{0i} + \eta_i s.$$

In all the corresponding transition probabilities, η_{0i} would be confounded
with the parameters τ_{ij} and therefore non-identifiable. This can be seen
as follows. With $\tau_{ii} = \eta_{0i} + \eta_i s$, γ_{ii} would then be given by

$$\gamma_{ii} = e^{\eta_i s} \big/ \left(e^{\eta_i s} + \textstyle\sum_{k \neq i} e^{\tau_{ik} - \eta_{0i}}\right),$$

Table 12.7 *Koeberg wind data (hourly): comparison of von Mises–HM models fitted to change in wind direction. The covariate, if any, is here used to influence the transition probabilities. The number of states is m, and the number of parameters k.*

covariate	m	k	$-l$	AIC	BIC
–	1	2	21821.350	43646.7	43663.6
–	2	6	8608.516	17229.0	17279.8
–	3	12	6983.205	13990.4	14092.0
–	4	20	6766.035	13572.1	13741.4
speed	2	8	6868.472	13752.9	13820.7
speed	3	15	5699.676	11429.4	11556.3
speed	4	24	5476.199	11000.4	11203.6
$\sqrt{\text{speed}}$	2	8	6771.533	13559.1	13626.8
$\sqrt{\text{speed}}$	3	15	5595.228	11220.5	11347.4
$\sqrt{\text{speed}}$	4	24	5361.759	**10771.5**	**10974.7**

and γ_{ij} (for $j \neq i$) by

$$\gamma_{ij} = e^{\tau_{ij} - \eta_{0i}} \Big/ \left(e^{\eta_i s} + \textstyle\sum_{k \neq i} e^{\tau_{ik} - \eta_{0i}} \right).$$

The results are displayed in Table 12.7, along with results for the corresponding models which use the square root of speed, rather than speed itself, as covariate. In these models, the square root of speed is in general more successful as a covariate than is speed.

12.3.4 Concentration parameter varying with lagged speed

A quite different way in which the square root of speed, lagged one hour, could be used as a covariate is via the concentration parameter (κ) of the von Mises distributions used here as the state-dependent distributions. The intuition underlying this proposal is that the higher the speed, the more concentrated will be the change in wind direction in the following hour (unless the state changes). One possibility is this: we assume that, given $S_{t-1} = s$, the concentration parameter in state i is

$$\log \kappa_i = \zeta_{i0} + \zeta_{i1}\sqrt{s}. \tag{12.2}$$

However, a model which seems both more successful (as judged by likelihood) and more stable numerically is this. Given $S_{t-1} = s$, let the concentration parameter in state i (not its logarithm) be a linear function of the square of the speed:

$$\kappa_i = \zeta_{i0} + \zeta_{i1}s^2. \tag{12.3}$$

Table 12.8 *Koeberg wind data (hourly): comparison of von Mises–HMMs fitted to change in wind direction. Here the covariate is used to influence the concentration parameter κ, as in Equation (12.3). The number of states is m, and the number of parameters k.*

covariate	m	k	$-l$	AIC	BIC
speed^2	1	3	10256.07	20518.1	20543.5
speed^2	2	8	4761.549	9539.1	9606.8
speed^2	3	15	4125.851	8281.7	8408.7
speed^2	4	24	3975.223	**7998.4**	**8201.6**

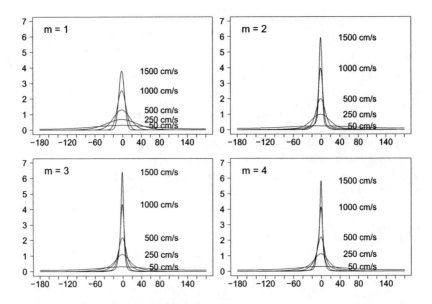

Figure 12.3 *Koeberg wind data (hourly), von Mises–HMMs of form (12.3): marginal distribution, for one to four states and several values of lagged speed, for change of direction (in degrees).*

To ensure that κ_i is positive we constrain ζ_{i0} and ζ_{i1} to be positive. Table 12.8 presents the log-likelihood values, AIC and BIC for four such models, and Figures 12.3 and 12.4 and Table 12.9 present some details of these models.

We have in this section considered a multiplicity of models for wind direction, but more variations yet would be possible. One could for in-

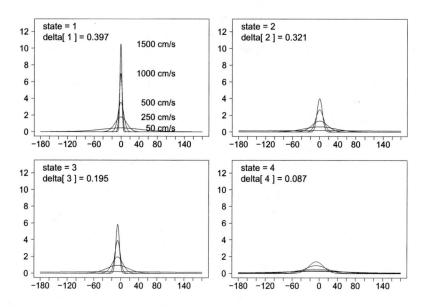

Figure 12.4 *Koeberg wind data (hourly), four-state von Mises–HMM of form (12.3): state-dependent distributions for change of direction (before multiplication by mixing probabilities δ_i).*

Table 12.9 *Koeberg wind data (hourly): parameters of four-state von Mises–HMM fitted to change in wind direction. Here the covariate is used to influence the concentration parameter κ, as in Equation (12.3).*

$$\mathbf{\Gamma} = \begin{pmatrix} 0.755 & 0.163 & 0.080 & 0.003 \\ 0.182 & 0.707 & 0.045 & 0.006 \\ 0.185 & 0.000 & 0.722 & 0.093 \\ 0.031 & 0.341 & 0.095 & 0.533 \end{pmatrix}$$

i	1	2	3	4
δ_i	0.397	0.321	0.195	0.087
μ_i	-0.0132	0.0037	-0.1273	-0.1179
ζ_{i0}	0.917	0.000	0.000	0.564
ζ_{i1}	31.01×10^{-5}	4.48×10^{-5}	9.61×10^{-5}	0.53×10^{-5}

stance allow the state-dependent location parameters μ_i (not the concentrations) of the change in direction to depend on the speed lagged one hour, or allow the concentrations to depend on the speed in different ways in the different states; there is no reason *a priori* why a single relation such as (12.3) should apply to all m states. In any application one would have to be guided very much by the intended use of the model.

Exercises

1.(a) Write an **R** function to generate a series of observations from an m-state categorical HMM with q categories, as described in Section 8.4.2. (Hint: modify the code in A.2.1.)

(b) Write functions to estimate the parameters of such a model. (Modify the code in A.1.)

(c) Generate a long series of observations using your code from 1(a), and estimate the parameters using your code from 1(b).

2. Use your code from 1(b) to fit two- and three-state models to the categorized wind directions data, and compare your models with those described in Section 12.2.1.

3. Generalize Exercise 1 to handle multinomial–HMMs, as described in Section 8.4.1, rather than merely categorical.

CHAPTER 13

Models for financial series

Because that's where the money is.

attributed to Willie Sutton

13.1 Financial series I: thinly traded shares on the Johannesburg Stock Exchange

One of the difficulties encountered in modelling the price series of shares listed on the Johannesburg Stock Exchange is that many of the shares are only thinly traded. The market is heavily dominated by institutional investors, and if for any reason a share happens not to be an 'institutional favourite' there will very likely be days, or even weeks, during which no trading of that share takes place. One approach is to model the presence or absence of trading quite separately from the modelling of the price achieved when trading does take place. This is analogous to the modelling of the sequence of wet and dry days separately from the modelling of the amounts of precipitation occurring on the wet days. It is therefore natural to consider, as models for the trading pattern of one or several shares, HMMs of the kind discussed by Zucchini and Guttorp (1991), who used them to represent the presence or absence of precipitation on successive days, at one or several sites.

In order to assess whether such models can be used successfully to represent trading patterns, data for six thinly traded shares were obtained from Dr D.C. Bowie, then of the University of Cape Town, and various models, including two-state HMMs, were fitted and compared. Of the six shares, three are from the coal sector and three from the diamonds sector. The coal shares are Amcoal, Vierfontein and Wankie, and the diamond shares Anamint, Broadacres and Carrigs. For all six shares the data cover the period from 5 October 1987 to 3 June 1991 (inclusive), during which time there were 910 days on which trading could take place. The data are therefore a multivariate binary time series of length 910.

13.1.1 Univariate models

The first two univariate models fitted to each of the six shares were a model assuming independence of successive observations and a Markov

Table 13.1 *Six thinly traded shares: minus log-likelihood values and BIC values achieved by five types of univariate model.*

Values of $-l$:

model	Amcoal	Vierf'n	Wankie	Anamint	Broadac	Carrigs
independence	543.51	629.04	385.53	612.03	599.81	626.88
Markov chain	540.89	611.07	384.57	582.64	585.76	570.25
second-order M. chain	539.89	606.86	383.87	576.99	580.06	555.67
2-state HMM, no trend	533.38	588.08	382.51	572.55	562.96	533.89
2-state HMM, single trend	528.07	577.51	381.28	562.55	556.21	533.88

Values of BIC:

model	Amcoal	Vierf'n	Wankie	Anamint	Broadac	Carrigs
independence	1093.83	1264.89	**777.88**	1230.88	1206.43	1260.58
Markov chain	1095.41	1235.77	782.77	1178.91	1185.15	1154.13
second-order M. chain	1107.03	1240.97	794.99	1181.23	1187.37	1138.59
2-state HMM, no trend	1094.01	1203.41	792.27	1172.35	1153.17	**1095.03**
2-state HMM, single trend	**1090.22**	**1189.08**	796.63	**1159.17**	**1146.49**	1101.83

chain (the latter fitted by conditional maximum likelihood). In all six cases, however, the sample ACF bore little resemblance to the ACF of the Markov chain model, and the Markov chain was therefore considered unsatisfactory. Two-state Bernoulli–HMMs, with and without time trend, and second-order Markov chains were also fitted, the second-order Markov chains by conditional maximum likelihood and the HMMs by unconditional. In the HMMs with trend, the probability of trading taking place on day t in state i is $_tp_i$, where

$$\text{logit } _tp_i = a_i + bt;$$

the trend parameter b is taken to be constant over states. Such a model has five parameters. In the HMM without trend, b is zero, and there are four parameters. The resulting log-likelihood and BIC values are shown in Table 13.1.

From that table we see that, of the five univariate models considered, the two-state HMM with a time trend fares best for four of the six shares: Amcoal, Vierfontein, Anamint and Broadacres. Of these four shares, all but Anamint show a negative trend in the probability of trading taking place, and Anamint a positive trend. In the case of Wankie, the model assuming independence of successive observations is chosen by BIC, and in the case of Carrigs an HMM without time trend is chosen.

Since a stationary HMM is chosen for Carrigs, it is interesting to compare the ACF of that model with the sample ACF and with the

Table 13.2 *Trading of Carrigs Diamonds: first eight terms of the sample ACF, compared with the autocorrelations of two possible models.*

ACF of Markov chain	0.350	0.122	0.043	0.015	0.005	0.002	0.001	0.000
sample ACF		0.349	0.271	0.281	0.237	0.230	0.202	0.177	0.200
ACF of HM model		0.321	0.293	0.267	0.244	0.223	0.203	0.186	0.169

ACF of the competing Markov chain model. For the HMM the ACF is $\rho(k) = 0.3517 \times 0.9127^k$, and for the Markov chain it is $\rho(k) = 0.3499^k$. Table 13.2 displays the first eight terms in each case. It is clear that the HMM comes much closer to matching the sample ACF than does the Markov chain model; a two-state HMM can model slow decay in $\rho(k)$ from any starting value $\rho(1)$, but a two-state Markov chain cannot.

13.1.2 Multivariate models

Two-state multivariate HMMs of two kinds were then fitted to each of the two groups of three shares: a model without time trend, and one which has a single (logit-linear) time trend common to the two states

Table 13.3 *Comparison of several multivariate models for the three coal shares and the three diamond shares.*

Coal shares

model	k	$-l$	BIC
3 'independence' models	3	1558.08	3136.60
3 univariate HMMs, no trend	12	1503.97	3089.69
3 univariate HMMs with trend	15	1486.86	**3075.93**
multivariate HMM, no trend	8	1554.01	3162.52
multivariate HMM, single trend	9	1538.14	3137.60

Diamond shares

model	k	$-l$	BIC
3 'independence' models	3	1838.72	3697.88
3 univariate HMMs, no trend	12	1669.40	3420.56
3 univariate HMMs with trend	15	1652.64	3407.48
multivariate HMM, no trend	8	1590.63	3235.77
multivariate HMM, single trend	9	1543.95	**3149.22**

Table 13.4 *Coal shares: univariate HMMs with trend.*

share	t.p.m.		a_1	a_2	b
Amcoal	$\begin{pmatrix} 0.774 & 0.226 \\ 0.019 & 0.981 \end{pmatrix}$		-0.332	1.826	-0.001488
Vierfontein	$\begin{pmatrix} 0.980 & 0.020 \\ 0.091 & 0.909 \end{pmatrix}$		0.606	3.358	-0.001792
Wankie	$\begin{pmatrix} 0.807 & 0.193 \\ 0.096 & 0.904 \end{pmatrix}$		-5.028	-0.943	-0.000681

and to the three shares in the group. The first type of model has eight parameters, the second has nine. These models were then compared with each other and with the 'product models' obtained by combining independent univariate models for the individual shares. The three types of product model considered were those based on independence of successive observations and those obtained by using the univariate HMMs with and without trend. The results are displayed in Table 13.3.

It is clear that, for the coal shares, the multivariate modelling has not been a success; the model consisting of three independent univariate hidden Markov models with trend is 'best'. We therefore give these three univariate models in Table 13.4. In each of these models $_tp_i$ is the probability that the relevant share is traded on day t if the state of the underlying Markov chain is i, and logit $_tp_i = a_i + bt$.

For the diamond shares, the best model of those considered is the multivariate HMM with trend. In this model logit $_tp_{ij} = a_{ij} + bt$, where $_tp_{ij}$ is the probability that share j is traded on day t if the state is i. The transition probability matrix is

$$\begin{pmatrix} 0.998 & 0.002 \\ 0.001 & 0.999 \end{pmatrix},$$

the trend parameter b is -0.003160, and the other parameters a_{ij} are as follows:

share	a_{1j}	a_{2j}
Anamint	1.756	1.647
Broadacres	0.364	0.694
Carrigs	1.920	-0.965

Table 13.5 *Coal shares: means, medians and standard deviations of bootstrap sample of estimators of parameters of two-state HMMs with time trend.*

share		$\hat{\gamma}_{12}$	$\hat{\gamma}_{21}$	$\hat{a}_1$	$\hat{a}_2$	$\hat{b}$
Amcoal	mean	0.251	0.048	-1.40	2.61	-0.00180
	median	0.228	0.024	-0.20	1.95	-0.00163
	s.d.	0.154	0.063	4.95	3.18	0.00108
Vierfontein	mean	0.023	0.102	0.599	4.06	-0.00187
	median	0.020	0.097	0.624	3.39	-0.00190
	s.d.	0.015	0.046	0.204	3.70	0.00041
Wankie	mean	0.145	0.148	-14.3	0.09	-0.00081
	median	0.141	0.089	-20.9	-0.87	-0.00074
	s.d.	0.085	0.167	10.3	4.17	0.00070

13.1.3 Discussion

The parametric bootstrap, with a sample size of 100, was used to investigate the distribution of the estimators in the models for the three coal shares which are displayed in Table 13.4. In this case the estimators show much more variability than do the estimators of the HMM used for the binary version of the geyser eruptions series; see p. 146. Table 13.5 gives for each of the three coal shares the bootstrap sample means, medians and standard deviations for the estimators of the five parameters. It will be noted that the estimators of a_1 and a_2 seem particularly variable. It is, however, true that, except in the middle of the range, very large differences on a logit scale correspond to small ones on a probability scale; two models with very different values of a_1 (for instance) may therefore produce almost identical distributions for the observations. For all three shares the trend parameter b seems to be more reliably estimated than the other parameters. If one is interested in particular in whether trading is becoming more or less frequent, this will be the parameter of most substantive interest.

As regards the multivariate HMM for the three diamond shares, it is perhaps surprising that the model is so much improved by the inclusion of a single (negative) trend parameter. In the corresponding univariate models the time trend was positive for one share, negative and of similar magnitude for another share, and negligible for the remaining share. Another criticism to which this multivariate model is open is that the off-diagonal elements of the transition probability matrix of its underlying Markov chain are so close to zero as to be almost negligible; on average only one or two changes of state would take place during a se-

quence of 910 observations. Furthermore, it is not possible to interpret
the two states as conditions in which trading is in general more likely
and conditions in which it is in general less so. This is because the probability of trading $({}_t p_{ij})$ is not consistently higher for one state i than the
other.

In view of the relative lack of success of multivariate HMMs in this
application, these models are not pursued here. The above discussion
does, we hope, serve as an illustration of the methodology, and suggests
that such multivariate models are potentially useful in studies of this
sort. They could, for instance, be used to model occurrences other than
the presence or absence of trading, e.g. the price (or volume) rising above
a given level.

13.2 Financial series II: a multivariate normal–HMM for returns on four shares

It is in practice often the case that a time series of share returns displays
kurtosis in excess of 3 and little or no autocorrelation, even though the
series of absolute or squared returns does display autocorrelation. These
— and some other — phenomena are so common that they are termed
'stylized facts'; for discussion see for instance Rydén, Teräsvirta and
Åsbrink (1998) or Bulla and Bulla (2007).

The model we shall discuss here is that the daily returns on four
shares have a multivariate normal distribution selected from one of m
such distributions by the underlying Markov chain, and that, conditional
on the underlying state, the returns on a share at a given time are
independent of those at any other time; that is, we assume longitudinal
conditional independence. We do not, however, assume contemporaneous
conditional independence. Indeed we impose here no structure on the
m 4×4 variance-covariance matrices, one for each possible state, nor
on the transition probability matrix $\mathbf{\Gamma}$. A model for p shares has in
all $m\{m - 1 + p(p + 3)/2\}$ parameters: $m^2 - m$ to determine $\mathbf{\Gamma}$, mp
state-dependent means, and $mp(p + 1)/2$ state-dependent variances and
covariances.

We evaluate the (log)-likelihood in the usual way and use nlm to
minimize minus the log-likelihood, but in order to do so we need an
unconstrained parametrization of the model. This is by now routine
in respect of the transition probabilities (e.g. via the generalized logit
transform). In addition, each of the m variance-covariance matrices $\mathbf{\Sigma}$ is
parametrized in terms of $\mathbf{T}$, its unique Cholesky upper-triangular 'square
root' such that the diagonal elements of $\mathbf{T}$ are positive; the relation between $\mathbf{\Sigma}$ and $\mathbf{T}$ is that $\mathbf{\Sigma} = \mathbf{T}'\mathbf{T}$.

We fitted models with two and three states to 500 consecutive daily

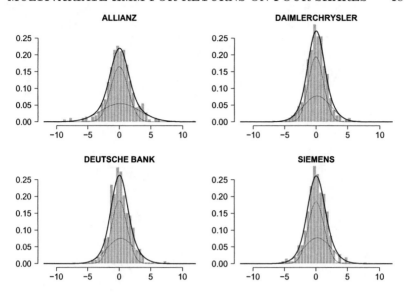

Figure 13.1 *Two-state HMM for four share returns: marginal distributions compared with histograms of the returns. The state-dependent distributions, multiplied by their mixing probabilities, are also shown.*

returns on the following four shares: Allianz (ALV), Deutsche Bank (DBK), DaimlerChrysler (DCX), and Siemens (SIE). These are all major components of the DAX30 index, with weights (as of 31 March 2006), of 8.71%, 7.55%, 6.97% and 10.04%. The 501 trading days used were 4 March 2003 to 17 February 2005, both inclusive. We computed the daily returns as $100 \log(s_t/s_{t-1})$, where s_t is the price on day t.

Of these two models the one that fitted better by BIC (but not AIC) was that with two states. In that case $-l = 3255.157$, there are 30 parameters, AIC $= 6570.3$ and BIC $= 6696.8$. For the three-state model the corresponding figures are $-l = 3208.915$, 48 parameters, AIC $= 6513.8$ and BIC $= 6716.1$. The details of the two-state model, which is depicted in Figure 13.1, are as follows.

$$\mathbf{\Gamma} = \begin{pmatrix} 0.918 & 0.082 \\ 0.059 & 0.941 \end{pmatrix} \qquad \boldsymbol{\delta} = (0.418, 0.582)$$

state-dependent means:

	ALV	DBK	DCX	SIE
state 1	0.172	0.154	0.252	0.212
state 2	0.034	−0.021	0.021	0.011

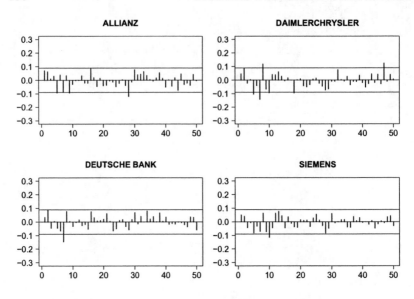

Figure 13.2 *Four share returns: sample ACF of returns.*

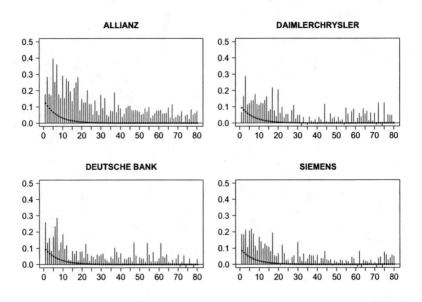

Figure 13.3 *Four share returns: sample ACF of squared returns, plus ACF of squared returns for two-state HMM (smooth curve).*

Σ for states 1 and 2 respectively:

$$\begin{pmatrix} 9.162 & 3.981 & 4.496 & 4.542 \\ 3.981 & 4.602 & 2.947 & 3.153 \\ 4.496 & 2.947 & 4.819 & 3.317 \\ 4.542 & 3.153 & 3.317 & 4.625 \end{pmatrix} \quad \begin{pmatrix} 1.985 & 1.250 & 1.273 & 1.373 \\ 1.250 & 1.425 & 1.046 & 1.085 \\ 1.273 & 1.046 & 1.527 & 1.150 \\ 1.373 & 1.085 & 1.150 & 1.590 \end{pmatrix}$$

standard deviations:

	ALV	DBK	DCX	SIE
state 1	3.027	2.145	2.195	2.151
state 2	1.409	1.194	1.236	1.261

correlation matrices for states 1 and 2 respectively:

$$\begin{pmatrix} 1.000 & 0.613 & 0.677 & 0.698 \\ 0.613 & 1.000 & 0.626 & 0.683 \\ 0.677 & 0.626 & 1.000 & 0.703 \\ 0.698 & 0.683 & 0.703 & 1.000 \end{pmatrix} \quad \begin{pmatrix} 1.000 & 0.743 & 0.731 & 0.773 \\ 0.743 & 1.000 & 0.709 & 0.721 \\ 0.731 & 0.709 & 1.000 & 0.738 \\ 0.773 & 0.721 & 0.738 & 1.000 \end{pmatrix}$$

One notable feature of this model is the clear ordering by volatility that emerges for these shares. The two states are one of high and one of low volatility, for all four shares. There was also a clear ordering by volatility in the three-state model; the states were of high, intermediate and low volatility, i.e. for all four shares. For a different set of shares, however, this ordering might well not emerge.

A feature of the two-state model, but not the three-state model, is that the ranges of correlations within each state are narrow and non-overlapping: 0.61–0.70 in state 1 (the more volatile state), and 0.71–0.77 in state 2.

We now compare properties of the two-state model with the corresponding properties of the observations. For all four shares, means and standard deviations agree extremely well. The same is not true of kurtosis. For the model the kurtoses of the four shares are 4.51 3.97 3.94, and 3.82, and for the observations the figures are 6.19, 4.69, 5.43, and 4.81 respectively. Although the model gives kurtosis well above 3 for each of the shares, the sample kurtosis is in each case much higher. However, much of this discrepancy is due to a few outlying returns in excess of 6% in absolute value, especially at the start of the period. The kurtosis of Allianz, in particular, is much reduced if one caps absolute returns at 6%.

We also compared sample and model autocorrelations of returns, auto-correlations of squared returns, and cross-correlations at lag zero. These cross-correlations match extremely well, and the ACFs of returns well (in that the model autocorrelations are very low and those of the sample negligible): see Figure 13.2. But the ACFs of squared returns do not match well: see Figure 13.3. This is similar to the finding of Rydén, Teräsvirta and Åsbrink (1998) that HMMs could not capture the sample ACF behaviour of absolute returns. Bulla and Bulla (2007) discuss the use of hidden *semi*-Markov models rather than HMMs in order to represent the squared-return behaviour better.

13.3 Financial series III: discrete state-space stochastic volatility models

Stochastic volatility (SV) models have attracted much attention in the finance literature; see e.g. Shephard (1996) and Omori *et al.* (2007). The first form of the model which we consider here, and the best-known, is the (Gaussian) SV model without leverage.

13.3.1 Stochastic volatility models without leverage

In such a model, the returns x_t on an asset ($t = 1, 2, \ldots, T$) satisfy

$$x_t = \varepsilon_t \beta \exp(g_t/2), \qquad g_{t+1} = \phi g_t + \eta_t, \qquad (13.1)$$

where $|\phi| < 1$ and $\{\varepsilon_t\}$ and $\{\eta_t\}$ are independent sequences of independent normal random variables, with ε_t standard normal and $\eta_t \sim N(0, \sigma^2)$. (This is the 'alternative parametrization' in terms of β indicated by Shephard (1996, p. 22), although we use ϕ and σ where Shephard uses γ_1 and σ_η.) We shall later allow ε_t and η_t to be dependent, in which case the model will be said to accommodate leverage; for the moment we exclude that possibility. This model has three parameters, β, ϕ and σ, and for identifiability reasons we constrain β to be positive. The model is both simple, in that it has only three parameters, and plausible in its structure, and some properties of the model are straightforward to establish. For instance, if $\{g_t\}$ is stationary, it follows that $g_t \sim N(0, \sigma^2/(1 - \phi^2))$. We shall assume that $\{g_t\}$ is indeed stationary.

But the principal difficulty in implementing the model in practice has been that it does not seem possible to evaluate the likelihood of the observations directly. Much ingenuity has been applied in the derivation and application of (*inter alia*) MCMC methods of estimating the parameters even in the case in which ε_t and η_t are assumed to be independent. By 1996 there was already a 'vast literature on fitting SV models' (Shephard, 1996, p. 35).

The dependence structure of such a model is precisely that of an HMM, but, unlike the models we have discussed so far, the underlying Markov process $\{g_t\}$ — essentially the log-volatility — is continuous-valued, not discrete. What we describe here, as an alternative to the estimation methods described, e.g., by Shephard (1996) and Kim, Shephard and

Chib (1998), is that the state-space of $\{g_t\}$ should be discretized into a sufficiently large number of states to provide a good approximation to continuity, and the well-established techniques and tractability of HMMs exploited in order to estimate the parameters of the resulting model. In particular, the ease of computation of the HMM likelihood enables one to fit models by numerical maximization of the likelihood, and to compute forecast distributions. The transition probability matrix of the Markov chain is structured here in such a way that an increase in the number of states does not increase the number of parameters; only the three parameters already listed are used.

The likelihood of the observations $\mathbf{x}^{(T)}$ is given by the T-fold multiple integral

$$p(\mathbf{x}^{(T)}) = \int \cdots \int p(\mathbf{x}^{(T)}, \mathbf{g}^{(T)}) \, d\mathbf{g}^{(T)},$$

the integrand of which can be decomposed as

$$p(g_1) \prod_{t=2}^{T} p(g_t \mid g_{t-1}) \prod_{t=1}^{T} p(x_t \mid g_t)$$

$$= p(g_1) \, p(x_1 \mid g_1) \prod_{t=2}^{T} p(g_t \mid g_{t-1}) \, p(x_t \mid g_t),$$

where p is used as a general symbol for a density. By discretizing the range of g_t sufficiently finely, we can approximate this integral by a multiple sum, and then evaluate that sum recursively by the methods of HMMs.

In detail, we proceed as follows. Let the range of possible g_t-values be split into m intervals (b_{i-1}, b_i), $i = 1, 2, \ldots, m$. If $b_{i-1} < g_t \leq b_i$, we shall say that g_t is in state i and denote this, as usual, by $C_t = i$. We denote by g_i^* a representative point in (b_{i-1}, b_i), e.g. the midpoint. The resulting T-fold sum approximating the likelihood is

$$\sum_{i_1=1}^{m} \sum_{i_2=1}^{m} \cdots \sum_{i_T=1}^{m} \Pr(C_1 = i_1) \, n(x_1; 0, \beta^2 \exp(g_{i_1}^*))$$

$$\times \prod_{t=2}^{T} \Pr(C_t = i_t \mid C_{t-1} = i_{t-1}) \, n(x_t; 0, \beta^2 \exp(g_{i_t}^*)), \quad (13.2)$$

and the transition probability $\Pr(C_t = j \mid C_{t-1} = i)$ is approximated by

$$\gamma_{ij} = N(b_j; \phi g_i^*, \sigma^2) - N(b_{j-1}; \phi g_i^*, \sigma^2)$$

$$= \Phi((b_j - \phi g_i^*)/\sigma) - \Phi((b_{j-1} - \phi g_i^*)/\sigma).$$

Here $n(\bullet; \mu, \sigma^2)$ is used to denote a normal density with mean μ and variance σ^2, N the corresponding (cumulative) distribution function, and Φ the standard normal distribution function. We are in effect saying that, although g_t has (conditional) mean ϕg_{t-1}, we shall proceed as if that mean were $\phi \times$ the midpoint of the interval in which g_{t-1} falls. With the assumption that the approximating Markov chain $\{C_t\}$ is stationary, (13.2) therefore gives us the usual matrix expression for the likelihood:

$$\boldsymbol{\delta}\mathbf{P}(x_1)\boldsymbol{\Gamma}\mathbf{P}(x_2)\ldots\boldsymbol{\Gamma}\mathbf{P}(x_T)\mathbf{1}' = \boldsymbol{\delta}\boldsymbol{\Gamma}\mathbf{P}(x_1)\boldsymbol{\Gamma}\mathbf{P}(x_2)\ldots\boldsymbol{\Gamma}\mathbf{P}(x_T)\mathbf{1}',$$

$\boldsymbol{\delta} = \mathbf{1}(\mathbf{I} - \boldsymbol{\Gamma} + \mathbf{U})^{-1}$ being the stationary distribution implied by the t.p.m. $\boldsymbol{\Gamma} = (\gamma_{ij})$, and $\mathbf{P}(x_t)$ being the diagonal matrix with ith diagonal element equal to the normal density $n(x_t; 0, \beta^2 \exp(g_i^*))$.

It is then a routine matter to evaluate this approximate likelihood and maximize it with respect to the three parameters of the model, transformed to allow for the constraints $\beta > 0$, $|\phi| < 1$, and $\sigma > 0$. However, in practice one also has to decide what m is, what range of g_t-values to allow for (i.e. b_0 to b_m), whether the intervals (b_{i-1}, b_i) should (e.g.) be of equal length, and which value g_i^* to take as representative of (b_{i-1}, b_i). Of these decisions, the choice of m can be expected to influence the accuracy of the approximation most. In the applications we describe in Sections 13.3.2 and 13.3.4, we have used equally spaced intervals represented by their midpoints.

13.3.2 Application: FTSE 100 returns

Shephard (1996, p. 39, Table 1.5) presents *inter alia* SV models fitted to the daily returns on the FTSE 100 index for the period 2 April 1986 to 6 May 1994; the return on day t is calculated as $100 \log(s_t/s_{t-1})$, where s_t is the index value at the close of day t. Using the technique described above, with a range of values of m and with g_t-values from -2.5 to 2.5, we have fitted models to FTSE 100 returns for that period.

Table 13.6 summarizes our findings and shows that the parameter estimates are reasonably stable by $m = 50$. Although we cannot expect our results to correspond exactly with those of Shephard, it is notable that our value of β (approximately 0.866) is very different from his (-0.452). The other two parameters are of roughly the same magnitude. We conjecture that the label β in column 1 of Shephard's Table 1.5 is inconsistent with his use of that symbol on p. 22, and therefore inconsistent with our usage. (We tested our code using simulated series.) Provisionally, our explanation for the apparent discrepancy is simply that the parameter β in Shephard's Table 1.5 is not the same as our parameter β.

Table 13.6 *SV model without leverage fitted to FTSE 100 returns, 2 April 1986 to 6 May 1994, plus comparable model from Shephard (1996).*

m	β	ϕ	σ
20	0.866	0.964	0.144
50	0.866	0.963	0.160
100	0.866	0.963	0.162
500	0.866	0.963	0.163

	β	γ_1	σ_η
Shephard (simulated EM)	-0.452	0.945	0.212

13.3.3 Stochastic volatility models with leverage

In the SV model without leverage, as described above, there is no feedback from past returns to the (log-) volatility process. As noted by Cappé *et al.* (2005, p. 28), this may be considered unnatural. We therefore discuss now a second, more general, form of the model.

As before, the returns x_t on an asset ($t = 1, 2, \ldots, T$) satisfy

$$x_t = \varepsilon_t \beta \exp(g_t/2), \qquad g_{t+1} = \phi g_t + \eta_t, \qquad (13.3)$$

where $|\phi| < 1$, but now ε_t and η_t are permitted to be dependent. More specifically, for all t

$$\begin{pmatrix} \varepsilon_t \\ \eta_t \end{pmatrix} \sim \mathrm{N}(\mathbf{0}, \boldsymbol{\Sigma}) \qquad \text{with} \qquad \boldsymbol{\Sigma} = \begin{pmatrix} 1 & \rho\sigma \\ \rho\sigma & \sigma^2 \end{pmatrix},$$

and the vectors $\begin{pmatrix} \varepsilon_t \\ \eta_t \end{pmatrix}$ are assumed independent.

This model has four parameters: β, ϕ, σ and ρ. We constrain β to be positive. If one does so, the model is equivalent to the model (2) of Omori *et al.* (2007), via $\beta = \exp(\mu/2)$ and $g_t = h_t - \mu$. It is also — with notational differences — the discrete-time ASV1 model (2.2) of Yu (2005); note that our η_t corresponds to Yu's v_{t+1}. Yu contrasts the ASV1 specification with that of Jacquier, Polson and Rossi (2004), and concludes that the ASV1 version is preferable. The parameter ρ is said to measure leverage; it is expected to be negative, in order to accommodate an increase in volatility following a drop in returns. The structure of the model is conveniently represented by the directed graph in Figure 13.4.

The likelihood of the observations $\mathbf{x}^{(T)}$ is in this case also given by the multiple integral

$$\int \cdots \int p(\mathbf{x}^{(T)}, \mathbf{g}^{(T)}) \, \mathrm{d}\mathbf{g}^{(T)}, \qquad (13.4)$$

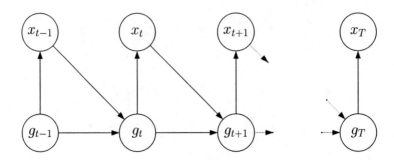

Figure 13.4 *Directed graph of stochastic volatility model with leverage.*

but here the integrand is decomposed as

$$p(g_1) \prod_{t=2}^{T} p(g_t \mid g_{t-1}, x_{t-1}) \prod_{t=1}^{T} p(x_t \mid g_t)$$

$$= p(g_1) p(x_1 \mid g_1) \prod_{t=2}^{T} p(g_t \mid g_{t-1}, x_{t-1}) p(x_t \mid g_t);$$

notice the dependence of $p(g_t \mid g_{t-1}, x_{t-1})$ on x_{t-1}.

We can approximate this integral as well by discretizing the range of g_t and evaluating the sum recursively. But in order to approximate this likelihood thus, we need the conditional distribution of g_{t+1} given g_t and x_t, or equivalently — since $x_t = \varepsilon_t \beta \exp(g_t/2)$ — given g_t and ε_t.

This is the distribution of η_t given ε_t, except that ϕg_t is added to the mean. The distribution of η_t given ε_t is $\mathrm{N}(\rho \sigma \varepsilon_t, \sigma^2 (1 - \rho^2))$; hence that of g_{t+1}, given g_t and ε_t, is

$$\mathrm{N}(\phi g_t + \rho \sigma \varepsilon_t, \sigma^2 (1 - \rho^2)).$$

Writing this distribution in terms of the observations x_t rather than the 'observation innovations' ε_t, we conclude that the required conditional distribution of g_{t+1} is

$$g_{t+1} \sim \mathrm{N}\left(\phi g_t + \frac{\rho \sigma x_t}{\beta \exp(g_t/2)}, \sigma^2 (1 - \rho^2) \right).$$

Hence $\Pr(g_t = j \mid g_{t-1} = i, x_{t-1} = x)$ is approximated by

$$\gamma_{ij}(x) = N(b_j; \mu(g_i^*, x), \sigma^2(1 - \rho^2)) - N(b_{j-1}; \mu(g_i^*, x), \sigma^2(1 - \rho^2))$$

$$= \Phi\left(\frac{b_j - \mu(g_i^*, x)}{\sigma\sqrt{1 - \rho^2}}\right) - \Phi\left(\frac{b_{j-1} - \mu(g_i^*, x)}{\sigma\sqrt{1 - \rho^2}}\right),$$

where we define

$$\mu(g_i^*, x) = \phi g_i^* + \frac{\rho \sigma x}{\beta \exp(g_i^*/2)}.$$

In this case, therefore, the approximate likelihood is

$$\delta \mathbf{P}(x_1)\mathbf{\Gamma}(x_1)\mathbf{P}(x_2)\mathbf{\Gamma}(x_2)\ldots\mathbf{\Gamma}(x_{T-1})\mathbf{P}(x_T)\mathbf{1}',$$

with δ here representing the distribution assumed for C_1, and $\mathbf{\Gamma}(x_t)$ the matrix with entries $\gamma_{ij}(x_t)$. This raises the question of what (if any) distribution for g_1 will produce stationarity in the process $\{g_t\}$, i.e. in the case of the model with leverage. In the case of the model without leverage, that distribution is normal with zero mean and variance $\sigma^2/(1 - \phi^2)$, and it is not unreasonable to conjecture that that may also be the case here.

Here we know that, given g_t and ε_t,

$$g_{t+1} \sim N(\phi g_t + \rho\sigma\varepsilon_t, \sigma^2(1 - \rho^2)).$$

That is,

$$g_{t+1} = \phi g_t + \rho\sigma\varepsilon_t + \sigma\sqrt{1 - \rho^2}Z,$$

where Z is independently standard normal. If it is assumed that $g_t \sim N(0, \sigma^2/(1 - \phi^2))$, it follows that g_{t+1} is (unconditionally) normal with mean zero and variance given by $\phi^2(\sigma^2/(1 - \phi^2)) + \rho^2\sigma^2 + \sigma^2(1 - \rho^2) = \sigma^2/(1 - \phi^2)$. In the 'with leverage' case also, therefore, the stationary distribution for $\{g_t\}$ is $N(0, \sigma^2/(1 - \phi^2))$, and that is the distribution we assume for g_1. The distribution we use for C_1 is that of g_1, discretized into the intervals (b_{i-1}, b_i).

13.3.4 Application: TOPIX returns

Table 13.7 *Summary statistics of TOPIX returns, calculated from opening prices 30 Dec. 1997 to 30 Dec. 2002, both inclusive.*

no. of returns	mean	std. dev.	max.	min.	+	−
1232	−0.02547	1.28394	5.37492	−5.68188	602	630

Using the daily opening prices of TOPIX (the Tokyo Stock Price Index) for the 1233 days from 30 December 1997 to 30 December 2002,

Table 13.8 *SV model with leverage fitted to TOPIX returns, opening prices 30 Dec. 1997 to 30 Dec. 2002, both inclusive, plus comparable figures from Table 5 of Omori* et al.

m	β	ϕ	σ	ρ
5	1.199	0.854	0.192	-0.609
10	1.206	0.935	0.129	-0.551
25	1.205	0.949	0.135	-0.399
50	1.205	0.949	0.140	-0.383
100	1.205	0.949	0.142	-0.379
200	1.205	0.949	0.142	-0.378

From Table 5 of Omori et al.:
posterior mean,

'unweighted'	1.2056	0.9511	0.1343	-0.3617
'weighted'	1.2052	0.9512	0.1341	-0.3578
95% interval	(1.089,1.318)	(0.908,0.980)	(0.091, 0.193)	$(-0.593, -0.107)$

Parametric bootstrap applied to the model with $m = 50$:
95% CI: (1.099, 1.293) (0.826, 0.973) (0.078, 0.262) $(-0.657, -0.050)$
correlations:

β		0.105	-0.171	0.004
ϕ			-0.752	-0.192
σ				0.324

both inclusive, we get a series of 1232 daily returns x_t with the summary statistics displayed in Table 13.7. This summary agrees completely with the statistics given by Omori *et al.* (2007) in their Table 4, although they state that they used closing prices. (We compute daily returns as $100\log(s_t/s_{t-1})$, where s_t is the price on day t, and we use the estimator with denominator T as the sample variance of T observations. The data were downloaded from `http://index.onvista.de` on 5 July 2006.)

Again using a range of values of m, and with g_t ranging from -2 to 2, we have fitted an SV model with leverage to these data. The results are summarized in Table 13.8. All parameter estimates are reasonably stable by $m = 50$ and, as expected, the estimate of the leverage parameter ρ is consistently negative. The results for $m = 50$ agree well with the two sets of point estimates ('unweighted' and 'weighted') presented by Omori *et al.* in their Table 5, and for all four parameters our estimate is close to the middle of their 95% interval.

We also applied the parametric bootstrap, with bootstrap sample size

500, to our model with $m = 50$, in order to estimate the (percentile) bootstrap 95% confidence limits, the standard errors and the correlations of our estimators. Some conclusions are as follows. The standard error of $\hat{\rho}$ (0.160) is relatively and absolutely the highest of the s.e.s, and there is a high negative correlation (-0.752) between the estimators of ϕ and σ. Overall, our conclusions are consistent with the corresponding figures in Table 5 of Omori *et al.*

13.3.5 Discussion

Alternative forms of feedback from the returns to the volatility process in an SV model are possible, and indeed have been proposed and implemented in a discrete state-space setting by Rossi and Gallo (2006). They attribute to Calvet and Fisher (2001, 2004) the first attempt to build accessible SV models based on high-dimensional regime switching. Fridman and Harris (1998) also present and implement a model involving such feedback; see their Equation (7). Their route to the evaluation, and hence maximization, of the likelihood is via recursive numerical evaluation of the multiple integral which gives the likelihood; see Equation (13.4) above.

A convenient feature of the model structure and estimation technique we have described is that the normality assumptions are not crucial. Firstly, one can use as the state process some Markov chain other than one based on a discretized Gaussian AR(1). Secondly, one can replace the normal distribution assumed for ε_t by some other distribution.

CHAPTER 14

Births at Edendale Hospital

14.1 Introduction

Haines, Munoz and van Gelderen (1989) have described the fitting of Gaussian ARIMA models to various discrete-valued time series related to births occurring during a 16-year period at Edendale Hospital in Natal*, South Africa. The data include monthly totals of mothers delivered and deliveries by various methods at the Obstetrics Unit of that hospital in the period from February 1970 to January 1986 inclusive. Although 16 years of data were available, Haines *et al.* considered only the final eight years' observations when modelling any series other than the total deliveries. This was because, for some of the series they modelled, the model structure was found not to be stable over the full 16-year period. In this analysis we have in general modelled only the last eight years' observations.

14.2 Models for the proportion Caesarean

One of the series considered by Haines *et al.*, to which they fitted two models, was the number of deliveries by Caesarean section. From their models they drew the conclusions (in respect of this particular series) that there is a clear dependence of present observations on past, and that there is a clear linear upward trend. In this section we describe the fitting of (discrete-valued) HM and Markov regression models to this series, Markov regression models, that is, in the sense in which that term is used by Zeger and Qaqish (1988). These models are of course rather different from those fitted by Haines *et al.* in that theirs, being based on the normal distribution, are continuous-valued. Furthermore, the discrete-valued models make it possible to model the proportion (as opposed to the number) of Caesareans performed in each month. Of the models proposed here, one type is 'observation-driven' and the other 'parameter-driven'; see Cox (1981) for these terms. The most important conclusion drawn from the discrete-valued models, and one which the Gaussian ARIMA models did not provide, is that there is a strong upward time trend in the proportion of the deliveries that are by Caesarean section.

---

* now KwaZulu–Natal

The two models that Haines *et al.* fitted to the time series of Caesareans performed, and that they found to fit very well, may be described as follows. Let Z_t denote the number of Caesareans in month t (February 1978 being month 1 and t running up to 96), and let the process $\{a_t\}$ be Gaussian white noise, i.e. uncorrelated random shocks distributed normally with zero mean and common variance σ_a^2. The first model fitted is the ARIMA(0,1,2) model with constant term:

$$\nabla Z_t = \mu + a_t - \theta_1 a_{t-1} - \theta_2 a_{t-2}. \qquad (14.1)$$

The maximum likelihood estimates of the parameters, with associated standard errors, are $\hat{\mu} = 1.02 \pm 0.39$, $\hat{\theta}_1 = 0.443 \pm 0.097$, $\hat{\theta}_2 = 0.393 \pm 0.097$ and $\hat{\sigma}_a^2 = 449.25$. The second model is an AR(1) with linear trend:

$$Z_t = \beta_0 + \beta_1 t + \phi Z_{t-1} + a_t, \qquad (14.2)$$

with parameter estimates as follows: $\hat{\beta}_0 = 120.2 \pm 8.2$, $\hat{\beta}_1 = 1.14 \pm 0.15$, $\hat{\phi} = 0.493 \pm 0.092$ and $\hat{\sigma}_a^2 = 426.52$.

Both of these models, (14.1) and (14.2), provide support for the conclusion of Haines *et al.* that there is a dependence of present observations on past, and a linear upward trend. Furthermore, the models are non-seasonal; the Box–Jenkins methodology used found no seasonality in the Caesareans series. The X-11-ARIMA seasonal adjustment method employed in an earlier study (Munoz, Haines and van Gelderen, 1987) did, however, find some evidence, albeit weak, of a seasonal pattern in the Caesareans series similar to a pattern that was observed in the 'total deliveries' series. (This latter series shows marked seasonality, with a peak in September, and in Haines *et al.* (1989) it is modelled by the seasonal ARIMA model $(0, 1, 1) \times (0, 1, 1)_{12}$.)

It is of some interest to model the proportion, rather than the number, of Caesareans in each month. (See Figure 14.1 for a plot of this proportion for the years 1978–1986.)

It could be the case, for instance, that any trend, dependence or seasonality apparently present in the number of Caesareans is largely inherited from the total deliveries, and a constant proportion Caesarean is an adequate model. On the other hand, it could be the case that there is an upward trend in the proportion of the deliveries that are by Caesarean and this accounts at least partially for the upward trend in the number of Caesareans. The two classes of model that we discuss in this section condition on the total number of deliveries in each month and seek to describe the principal features of the proportion Caesarean.

Now let n_t denote the total number of deliveries in month t. A very general possible model for $\{Z_t\}$ which allows for trend, dependence on previous observations and seasonality, in the proportion Caesarean, is as follows. Suppose that, conditional on the history $\mathbf{Z}^{(t-1)} = \{Z_s : s \leq t{-}1\}$,

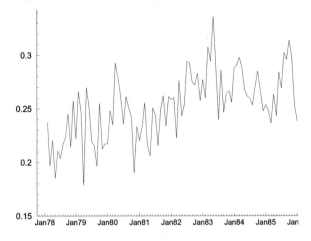

Figure 14.1 *Edendale births: monthly numbers of deliveries by Caesarean section, as a proportion of all deliveries, February 1978 – January 1986.*

Z_t is distributed binomially with parameters n_t and $_tp$, where, for some positive integer q,

$$\begin{aligned}
\text{logit}\,_tp \;=\; & \alpha_1 + \alpha_2 t + \beta_1(Z_{t-1}/n_{t-1}) + \beta_2(Z_{t-2}/n_{t-2}) + \cdots \\
& + \beta_q(Z_{t-q}/n_{t-q}) + \gamma_1 \sin(2\pi t/12) + \gamma_2 \cos(2\pi t/12).
\end{aligned} \tag{14.3}$$

This is, in the terminology of Zeger and Qaqish (1988), a Markov regression model, and generalizes the model described by Cox (1981) as an 'observation-driven linear logistic autoregression', in that it incorporates trend and seasonality and is based on a binomial rather than a Bernoulli distribution. It is observation-driven in the sense that the distribution of the observation at a given time is specified in terms of the observations at earlier times. Clearly it is possible to add further terms to the above expression for logit $_tp$ to allow for the effect of any further covariates thought relevant, e.g. the number or proportion of deliveries by various instrumental techniques.

It does not seem possible to formulate an unconditional maximum likelihood procedure to estimate the parameters α_1, α_2, β_1, ..., β_q, γ_1 and γ_2 of the model (14.3). It is, however, straightforward to compute estimates of these parameters by maximizing a conditional likelihood. If for instance no observations earlier than Z_{t-1} appear in the model, the product

$$\prod_{t=1}^{96} \binom{n_t}{z_t} {}_t p^{z_t} (1 - {}_t p)^{n_t - z_t}$$

is the likelihood of $\{Z_t : t = 1, \dots, 96\}$, conditional on Z_0. Maximization thereof with respect to $\alpha_1, \alpha_2, \beta_1, \gamma_1$ and γ_2 yields estimates of these parameters, and can be accomplished simply by performing a logistic regression of Z_t on t, Z_{t-1}/n_{t-1}, $\sin(2\pi t/12)$ and $\cos(2\pi t/12)$.

In the search for a suitable model the following explanatory variables were considered: t (i.e. the time in months, with February 1978 as month 1), t^2, the proportion Caesarean lagged one, two, three or twelve months, sinusoidal terms at the annual frequency (as in (14.3)), the calendar month, the proportion and number of deliveries by forceps or vacuum extraction, and the proportion and number of breech births. Model selection was performed by means of AIC and BIC. The **R** function used, glm, does not provide l, the log-likelihood, but it does provide the deviance, from which the log-likelihood can be computed (McCullagh and Nelder, 1989, p. 33). Here the maximum log-likelihood of a full model is -321.0402, from which it follows that $-l = 321.04 + \frac{1}{2} \times$ deviance.

The best models found with between one and four explanatory variables (other than the constant term) are listed in Table 14.1, as well as several other models that may be of interest. These other models are: the model with constant term only; the model with Z_{t-1}/n_{t-1}, the previous proportion Caesarean, as the only explanatory variable; and two models which replace the number of forceps deliveries as covariate by the proportion of instrumental deliveries. The last two models were included because the proportion of instrumental deliveries may seem a more sensible explanatory variable than the one it replaces; it will be observed, however, that the models involving the number of forceps deliveries are preferred by AIC and BIC. (By 'instrumental deliveries' we mean those which are either by forceps or by vacuum extraction.) Note, however, that both AIC and BIC indicate that one could still gain by inclusion of a fifth explanatory variable in the model.

The strongest conclusion we may draw from these models is that there is indeed a marked upward time trend in the proportion Caesarean. Secondly, there is positive dependence on the proportion Caesarean in the previous month. The negative association with the number (or proportion) of forceps deliveries is not surprising in view of the fact that delivery by Caesarean and by forceps are in some circumstances alternative techniques. As regards seasonality, the only possible seasonal pattern found in the proportion Caesarean is the positive 'October effect'. Among the calendar months only October stood out as having some explanatory power. As can be seen from Table 14.1, the indicator variable specify-

Table 14.1 *Edendale births: models fitted to the logit of the proportion Caesarean.*

explanatory variables	coefficients	deviance	AIC	BIC
constant	−1.253	208.92	855.00	860.13
t (time in months)	0.003439			
constant	−1.594	191.70	839.78	847.47
t	0.002372			
previous proportion Caesarean	1.554			
constant	−1.445	183.63	833.71	843.97
t	0.001536			
previous proportion Caesarean	1.409			
no. forceps deliveries in month t	−0.002208			
constant	−1.446	175.60	**827.68**	**840.50**
t	0.001422			
previous proportion Caesarean	1.431			
no. forceps deliveries in month t	−0.002393			
October indicator	0.08962			
constant	−1.073	324.05	968.13	970.69
constant	−1.813	224.89	870.97	876.10
previous proportion Caesarean	2.899			
constant	−1.505	188.32	838.40	848.66
t	0.002060			
previous proportion Caesarean	1.561			
proportion instrumental				
deliveries in month t	−0.7507			
constant	−1.528	182.36	834.44	847.26
t	0.002056			
previous proportion Caesarean	1.590			
proportion instrumental				
deliveries in month t	−0.6654			
October indicator	0.07721			

ing whether the month was October was included in the 'best' set of four explanatory variables. A possible reason for an October effect is as follows. An overdue mother is more likely than others to give birth by Caesarean, and the proportion of overdue mothers may well be highest in October because the peak in total deliveries occurs in September.

Since the main conclusion emerging from the above logit-linear models

Table 14.2 *Edendale births: the three two-state binomial–HMMs fitted to the proportion Caesarean. (The time in months is denoted by t, and February 1978 is month 1.)*

logit $_t p_i$	$-l$	AIC	BIC	γ_{12}	γ_{21}	α_1	α_2	β_1	β_2
α_i	420.322	848.6	858.9	0.059	0.086	-1.184	-0.960	–	–
$\alpha_i + \beta t$	402.350	**814.7**	**827.5**	0.162	0.262	-1.317	-1.140	0.003298	
$\alpha_i + \beta_i t$	402.314	816.6	832.0	0.161	0.257	-1.315	-1.150	0.003253	0.003473

is that there is a marked upward time trend in the proportion Caesarean, it is of interest also to fit HMMs with and without time trend. The HMMs we use in this application have two states and are defined as follows. Suppose $\{C_t\}$ is a stationary homogeneous Markov chain on state space $\{1, 2\}$, with transition probability matrix

$$\mathbf{\Gamma} = \begin{pmatrix} 1 - \gamma_{12} & \gamma_{12} \\ \gamma_{21} & 1 - \gamma_{21} \end{pmatrix}.$$

Suppose also that, conditional on the Markov chain, Z_t has a binomial distribution with parameters n_t and p_i, where $C_t = i$. A model without time trend assumes that p_1 and p_2 are constants and has four parameters. One possible model which allows p_i to depend on t has logit $_t p_i = \alpha_i + \beta t$ and has five parameters. A more general model yet, with six parameters, has logit $_t p_i = \alpha_i + \beta_i t$.

Maximization of the likelihood of the last eight years' observations gives the three models appearing in Table 14.2 along with their associated log-likelihood, AIC and BIC values. It may be seen that, of the three models, that with a single time-trend parameter and a total of five parameters achieves the smallest AIC and BIC values.

In detail, that model is as follows, t being the time in months and February 1978 being month 1:

$$\mathbf{\Gamma} = \begin{pmatrix} 0.838 & 0.162 \\ 0.262 & 0.738 \end{pmatrix},$$

$$\text{logit } _t p_1 = -1.317 + 0.003298t,$$

$$\text{logit } _t p_2 = -1.140 + 0.003298t.$$

The model can be described as consisting of a Markov chain with two fairly persistent states, along with their associated time-dependent probabilities of delivery being by Caesarean, the (upward) time trend being the same, on a logit scale, for the two states. State 1 is rather more likely than state 2, because the stationary distribution is $(0.618, 0.382)$, and has associated with it a lower probability of delivery being by Caesarean. For state 1 that probability increases from 0.212 in month 1 to 0.269 in month 96, and for state 2 from 0.243 to 0.305. The corresponding unconditional probability increases from 0.224 to 0.283. It may or may not be possible to interpret the states as (for instance) nonbusy and busy periods in the Obstetrics Unit of the hospital, but without further information, e.g. on staffing levels, such an interpretation would be speculative.

It is true, however, that other models can reasonably be considered. One possibility, suggested by inspection of Figure 14.1, is that the proportion Caesarean was constant until January 1981, then increased linearly to a new level in about January 1983. Although we do not pursue such a model here, it is possible to fit an HMM incorporating this feature. (Models with change-points are discussed and used in Chapter 15.)

If one wishes to use the chosen model to forecast the proportion Caesarean at time 97 for a given number of deliveries, what is needed is the one-step-ahead forecast distribution of Z_{97}, i.e. the distribution of Z_{97} conditional on $Z_1, \ldots, Z_{96}$. This is given by the likelihood of $Z_1, \ldots, Z_{97}$ divided by that of $Z_1, \ldots, Z_{96}$. More generally, the k-step-ahead forecast distribution, i.e. the conditional probability that $Z_{96+k} = z$, is given by a ratio of likelihoods, as described in Section 5.2.

The difference in likelihood between the HMMs with and without time trend is convincing evidence of an upward trend in the proportion Caesarean, and confirms the main conclusion drawn above from the logit-linear models. Although Haines *et al.* concluded that there is an upward trend in the number of Caesareans, it does not seem possible to draw any conclusion about the proportion Caesarean from their models, or from any other ARIMA models.

It is of interest also to compare the fit of the five-parameter HMM to the data with that of the logistic autoregressive models. Here it should be noted that the HMM produces a lower value of $-l$ (402.3) than does the logistic autoregressive model with four explanatory variables (408.8), without making use of the additional information used by the logistic autoregression. It does not use z_0, the number of Caesareans in January 1978, nor does it use information on forceps deliveries or the calendar month. It seems therefore that HMMs have considerable potential as simple yet flexible models for examining dependence on covariates (such as time) in the presence of serial dependence.

14.3 Models for the total number of deliveries

If one wishes to project the number of Caesareans, however, a model for the proportion Caesarean is not sufficient; one needs also a model for the

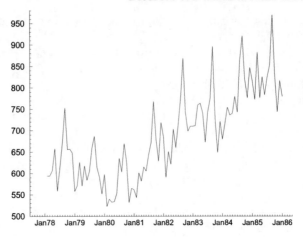

Figure 14.2 *Edendale births: monthly totals of deliveries, February 1978 –
January 1986.*

total number of deliveries, which is a series of unbounded counts. The
model of Haines *et al.* for the total deliveries was the seasonal ARIMA
model $(0, 1, 1) \times (0, 1, 1)_{12}$ without constant term. For this series (unlike
the others) they used all 16 years' data to fit the model, but we continue
here to model only the final eight years' observations.

First, five two-state Poisson–HMMs. were fitted to the monthly totals of deliveries (depicted in Figure 14.2). One was a model without
covariates, that is,

$$\log {}_t\lambda_i = a_i; \qquad (14.4)$$

and the other four models for $\log {}_t\lambda_i$ were of the following forms:

$$\log {}_t\lambda_i \;=\; a_i + bt; \qquad (14.5)$$
$$\log {}_t\lambda_i \;=\; a_i + bt + c\cos(2\pi t/12) + d\sin(2\pi t/12); \qquad (14.6)$$
$$\log {}_t\lambda_i \;=\; a_i + bt + c\cos(2\pi t/12) + d\sin(2\pi t/12) + fn_{t-1}; \qquad (14.7)$$
$$\log {}_t\lambda_i \;=\; a_i + bt + c\cos(2\pi t/12) + d\sin(2\pi t/12) + fn_{t-1} + gn_{t-2}. \qquad (14.8)$$

Models (14.7) and (14.8) are examples of the incorporation of extra
dependencies at observation level as described in Section 8.6; that is,
they do not assume conditional independence of the observations $\{n_t\}$
given the Markov chain. But this does not significantly complicate the
likelihood evaluation, as the state-dependent probabilities can just treat
n_{t-1} and n_{t-2} in the same way as any other covariate. The models fitted
are summarized in Table 14.3.

Table 14.3 *Edendale births: summary of the five two-state Poisson–HMMs fitted to the number of deliveries. (The time in months is denoted by t, and February 1978 is month 1.)*

$\log{}_t\lambda_i$	$-l$	AIC BIC	γ_{12} γ_{21}	a_1 a_2	b	c d	f	g
(14.4)	642.023	1292.0	0.109	6.414	–	–	–	–
		1302.3	0.125	6.659		–		
(14.5)	553.804	1117.6	0.138	6.250	0.004873	–	–	–
		1130.4	0.448	6.421		–		
(14.6)	523.057	1060.1	0.225	6.267	0.004217	−0.0538	–	–
		1078.1	0.329	6.401		−0.0536		
(14.7)	510.840	**1037.7**	0.428	5.945	0.002538	−0.0525	0.000586	–
		1058.2	0.543	6.069		−0.0229		
(14.8)	510.791	1039.6	0.415	5.939	0.002511	−0.0547	0.000561	0.0000365
		1062.7	0.537	6.063		−0.0221		

It may be seen that, of these five models, model (14.7), with a total of eight parameters, achieves the smallest AIC and BIC values. That model is as follows:

$$\log{}_t\lambda_i = 5.945/6.069 + 0.002538t - 0.05253\cos(2\pi t/12)$$
$$- 0.02287\sin(2\pi t/12) + 0.0005862n_{t-1}.$$

In passing, we may mention that models (14.5)–(14.8) were fitted by means of the **R** function `constrOptim` (for constrained optimization). It is possible to refine these models for the monthly totals by allowing for the fact that the months are of unequal length, and by allowing for seasonality of frequency higher than annual, but these variations were not pursued.

A number of log-linear models were then fitted by `glm`. Table 14.4 compares the models fitted by `glm` to the total deliveries, and from that table it can be seen that of these models BIC selects the model incorporating time trend, sinusoidal components at the annual frequency, and the number of deliveries in the previous month (n_{t-1}). The details of this model are as follows. Conditional on the history, the number of deliveries in month t (N_t) is distributed Poisson with mean ${}_t\lambda$, where

$$\log{}_t\lambda = 6.015 + 0.002436t - 0.03652\cos(2\pi t/12)$$
$$- 0.02164\sin(2\pi t/12) + 0.0005737n_{t-1}.$$

(Here, as before, February 1978 is month 1.)

Both the HMMs and the log-linear autoregressions reveal time trend,

Table 14.4 *Edendale births: models fitted by* `glm` *to the log of the mean number of deliveries.*

Explanatory variables	Deviance	$-l$	AIC	BIC
t	545.9778	674.3990	1352.8	1357.9
t, sinusoidal terms *	428.6953	615.7577	1239.5	1249.8
t, sinusoidal terms, n_{t-1}	356.0960	579.4581	1168.9	**1181.7**
t, sinusoidal terms, n_{t-1}, n_{t-2}	353.9086	578.3644	**1168.7**	1184.1
sinusoidal terms, n_{t-1}	464.8346	633.8274	1275.7	1285.9
t, n_{t-1}	410.6908	606.7555	1219.5	1227.2
n_{t-1}	499.4742	651.1472	1306.3	1311.4

* This is a one-state model similar to the HMM (14.6), and its log-likelihood value can be used to some extent to check the code for the HMM. Similar comments apply to the other models listed above.

seasonality, and dependence on the number of deliveries in the previous month. On the basis of both AIC and BIC the Poisson–HMM (14.7) is the best model.

14.4 Conclusion

The conclusion is therefore twofold. If a model for the number of Caesareans, given the total number of deliveries, is needed, the binomial–HMM with time trend is best of all of those considered (including various logit-linear autoregressive models). If a model for the total deliveries is needed (e.g. as a building-block in projecting the number of Caesareans), then the Poisson–HMM with time trend, 12-month seasonality and dependence on the number in the previous month is the best of those considered — contrary to our conclusion in MacDonald and Zucchini (1997), which was unduly pessimistic about the Poisson–HMMs.

The models chosen here suggest that there is a clear upward time trend in both the total deliveries and the proportion Caesarean, and seasonality in the total deliveries.

CHAPTER 15

Homicides and suicides in Cape Town, 1986–1991

15.1 Introduction

In South Africa, as in the USA, gun control is a subject of much public interest and debate. In a project intended to study the apparently increasing tendency for violent crime to involve firearms, Dr L.B. Lerer collected data relating to homicides and suicides from the South African Police mortuary in Salt River, Cape Town. Records relating to most of the homicide and suicide cases occurring in metropolitan Cape Town are kept at this mortuary. The remaining cases are dealt with at the Tygerberg Hospital mortuary. It is believed, however, that the exclusion of the Tygerberg data does not materially affect the conclusions.

The data consist of all the homicide and suicide cases appearing in the deaths registers relating to the six-year period from 1 January 1986 to 31 December 1991. In each such case the information recorded included the date and cause of death. The five (mutually exclusive) categories used for the cause of death were: firearm homicide, nonfirearm homicide, firearm suicide, nonfirearm suicide, and 'legal intervention homicide'. (This last category refers to homicide by members of the police or army in the course of their work. In what follows, the word homicide, if unqualified, means homicide other than that resulting from such legal intervention.) Clearly some of the information recorded in the deaths registers could be inaccurate, e.g. a homicide recorded as a suicide, or a legal intervention homicide recorded as belonging to another category. This has to be borne in mind in drawing conclusions from the data.

15.2 Firearm homicides as a proportion of all homicides, suicides and legal intervention homicides

One question of interest that was examined by means of HMMs was whether there is an upward trend in the proportion of all the deaths recorded that are firearm homicides. This is of course quite distinct from the question of whether there is an upward trend in the *number* of firearm homicides. The latter kind of trend could be caused by an increase in the population exposed to risk of death, without there being any other

209

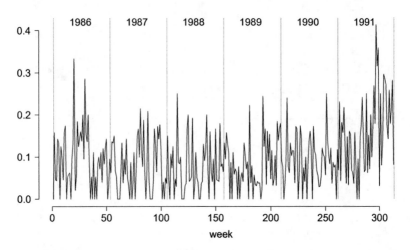

Figure 15.1 *Firearm homicides 1986–1991, as a proportion of all homicides, suicides and legal intervention homicides.*

relevant change. This distinction is important because of the rapid urbanization which has taken place in South Africa and has caused the population in and around Cape Town to increase dramatically.

Four models were fitted to the 313 weekly totals of firearm homicides (given the weekly totals of all the deaths recorded); for a plot of the firearm homicides in each week, as a proportion of all homicides, suicides and legal intervention homicides, see Figure 15.1.

The four models are: a two-state binomial–HMM with constant 'success probabilities' p_1 and p_2; a similar model with a linear time trend (the same for both states) in the logits of those probabilities; a model allowing differing time-trend parameters in the two states; and finally, a model which assumes that the success probabilities are piecewise constant with a single change-point at time 287 — i.e. on 2 July 1991, 26 weeks before the end of the six-year period studied. The time of the change-point was chosen because of the known upsurge of violence in some of the areas adjacent to Cape Town, in the second half of 1991. Much of this violence was associated with the 'taxi wars', a dispute between rival groups of public transport operators.

The models were compared on the basis of AIC and BIC. The results are shown in Table 15.1. Broadly, the conclusion from BIC is that a single (upward) time trend is better than either no trend or two trend parameters, but the model with a change-point is the best of the four. The details of this model are as follows. The underlying Markov chain

Table 15.1 *Comparison of several binomial–HMMs fitted to the weekly counts of firearm homicides, given the weekly counts of all homicides, suicides and legal intervention homicides.*

model with	k	$-l$	AIC	BIC
p_1 and p_2 constant	4	590.258	1188.5	1203.5
one time-trend parameter	5	584.337	1178.7	1197.4
two time-trend parameters	6	579.757	1171.5	1194.0
change-point at time 287	6	573.275	**1158.5**	**1181.0**

has transition probability matrix

$$\left(\begin{array}{cc} 0.658 & 0.342 \\ 0.254 & 0.746 \end{array} \right)$$

and stationary distribution (0.426, 0.574). The probabilities p_1 and p_2 are given by (0.050, 0.116) for weeks 1–287, and by (0.117, 0.253) for weeks 288–313. From this it appears that the proportion of the deaths that are firearm homicides was substantially greater by the second half of 1991 than it was earlier, and that this change is better accommodated by a discrete shift in the probabilities p_1 and p_2 than by gradual movement with time, at least gradual movement of the kind that we have incorporated into the models with time trend. (In passing, this use of a discrete shift further illustrates the flexibility of HMMs.) One additional model was fitted: a model with change-point at the end of week 214. That week included 2 February 1990, on which day President de Klerk made a speech which is widely regarded as a watershed in South Africa's history. That model yielded a log-likelihood of -579.83, and AIC and BIC values of 1171.67 and 1194.14. Such a model is therefore inferior to the model with the change-point at the end of week 287.

15.3 The number of firearm homicides

In order to model the number (rather than the proportion) of firearm homicides, Poisson–HMMs were also fitted. The weekly counts of firearm homicides are shown in Figure 15.2. There is a marked increase in the level of the series at about week 287 (mid-1991), but another, less distinct, pattern is discernible in the values prior to that date. There seems to be persistence in those periods when the counts are high (e.g. around weeks 25 and 200); runs of relatively calm weeks seem to alternate with runs of increased violence. This observation suggests that a two-state HMM might be an appropriate model.

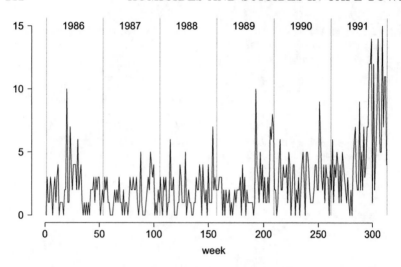

Figure 15.2 *Weekly counts of firearm homicides, 1986–1991.*

Table 15.2 *Comparison of several Poisson–HMMs fitted to weekly counts of firearm homicides.*

model with	k	−l	AIC	BIC
λ_1 and λ_2 constant	4	626.637	1261.3	1276.3
loglinear trend	5	606.824	1223.6	1242.4
log-quadratic trend	6	602.273	**1216.5**	**1239.0**
change-point at time 287	6	605.559	1223.1	1245.6

The four models fitted in this case were: a two-state model with constant conditional means λ_1 and λ_2; a similar model with a single linear trend in the logs of those means; a model with a quadratic trend therein; and finally, a model allowing for a change-point at time 287. A comparison of these models is shown in Table 15.2.

The conclusion is that, of the four models, the model with a quadratic trend in the conditional means is best. In detail, that model is as follows. The underlying Markov chain has transition probability matrix given by $\begin{pmatrix} 0.881 & 0.119 \\ 0.416 & 0.584 \end{pmatrix}$ and stationary distribution $(0.777, 0.223)$. The conditional means are given by

$$\log {}_t\lambda_i = 0.4770/1.370 - 0.004858t + 0.00002665t^2,$$

where t is the week number and i the state. The fact that a smooth trend

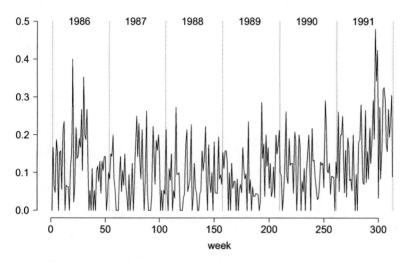

Figure 15.3 *Firearm homicides as a proportion of all homicides, 1986–1991. The week ending on 2 July 1991 is week 287.*

works better here than does a discrete shift may possibly be explained by population increase due to migration, especially towards the end of the six-year period.

15.4 Firearm homicides as a proportion of all homicides, and firearm suicides as a proportion of all suicides

A question of interest that arises from the apparently increased proportion of firearm homicides is whether there is any similar tendency in respect of suicides. Here the most interesting comparison is between firearm homicides as a proportion of all homicides and firearm suicides as a proportion of all suicides. Plots of these two proportions appear as Figures 15.3 and 15.4. Binomial–HMMs of several types were used to model these proportions, and the results are given in Tables 15.3 and 15.4.

The chosen models for these two proportions are as follows. For the firearm homicides the Markov chain has transition probability matrix $\begin{pmatrix} 0.695 & 0.305 \\ 0.283 & 0.717 \end{pmatrix}$ and stationary distribution (0.481, 0.519). The probabilities p_1 and p_2 are given by (0.060, 0.140) for weeks 1–287, and by (0.143, 0.283) for weeks 288–313. The unconditional probability that a homicide involved the use of a firearm is therefore 0.102 before the change-point, and 0.216 thereafter. For the firearm suicides, the transition probability matrix is $\begin{pmatrix} 0.854 & 0.146 \\ 0.117 & 0.883 \end{pmatrix}$, and the stationary distri-

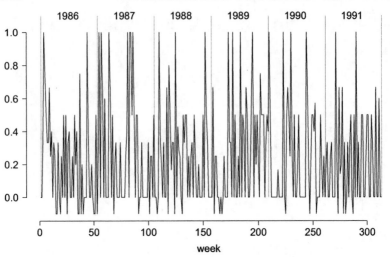

Figure 15.4 *Firearm suicides as a proportion of all suicides, 1986–1991. (Negative values indicate no suicides in that week.)*

Table 15.3 *Comparison of binomial–HMMs for firearm homicides given all homicides.*

model with	k	$-l$	AIC	BIC
p_1 and p_2 constant	4	590.747	1189.5	1204.5
one time-trend parameter	5	585.587	1181.2	1199.9
two time-trend parameters	6	580.779	1173.6	1196.0
change-point at time 287	6	575.037	**1162.1**	**1184.6**

bution is (0.446, 0.554). The probabilities p_1 and p_2 are given by (0.186, 0.333), and the unconditional probability that a suicide involves a firearm is 0.267.

A question worth considering, however, is whether time series models are needed at all for these proportions. Is it not perhaps sufficient to fit a one-state model, i.e. a model which assumes independence of the consecutive observations but is otherwise identical to one of the time series models described above? Models of this type were therefore fitted both to the firearm homicides as a proportion of all homicides, and to the firearm suicides as a proportion of all suicides. The comparisons of models are presented in Tables 15.5 and 15.6, in which the parameter p represents the probability that a death involves a firearm.

The conclusions that may be drawn from these models are as follows.

Table 15.4 *Comparison of binomial–HMMs for firearm suicides given all suicides.*

model with	k	$-l$	AIC	BIC
p_1 and p_2 constant	4	289.929	**587.9**	**602.8**
one time-trend parameter	5	289.224	588.4	607.2
two time-trend parameters	6	288.516	589.0	611.5
change-point at time 287	6	289.212	590.4	612.9

Table 15.5 *Comparison of several 'independence' models for firearm homicides as a proportion of all homicides.*

model with	k	$-l$	AIC	BIC
p constant	1	637.458	1276.9	1280.7
time trend in p	2	617.796	1239.6	1247.1
change-point at time 287	2	590.597	**1185.2**	**1192.7**

For the homicides, the models based on independence are without exception clearly inferior to the corresponding HM time series models. There is sufficient serial dependence present in the proportion of the homicides involving a firearm to render inappropriate any analysis based on an assumption of independence. For the suicides the situation is reversed; the models based on independence are in general superior. There is in this case no convincing evidence of serial dependence, and time series models do not appear to be necessary. The 'best' model based on independence assigns a value of 0.268 (= 223/833) to the probability that a suicide involves the use of a firearm, which is of course quite consistent with the value (0.267) implied by the chosen HMM.

To summarize the conclusions, therefore, we may say that the pro-

Table 15.6 *Comparison of several 'independence' models for firearm suicides as a proportion of all suicides.*

model with	k	$-l$	AIC	BIC
p constant	1	291.166	**584.3**	**588.1**
time trend in p	2	290.275	584.6	592.0
change-point at time 287	2	291.044	586.1	593.6

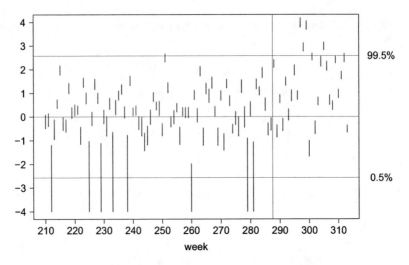

Figure 15.5 *Firearm homicides as a proportion of all homicides: forecast pseudo-residual segments computed from a model for weeks 1–209 only. The vertical line is immediately after week 287.*

portion of homicides that involve firearms does indeed seem to be at a higher level after June 1991, but that there is no evidence of a similar upward shift (or trend) in respect of the proportion of the suicides that involve firearms. There is evidence of serial dependence in the proportion of homicides that involve firearms, but not in the corresponding proportion of suicides.

In view of the finding that the proportion of homicides due to firearms seems to be higher after June 1991, it is interesting to see whether the monitoring technique introduced in Section 6.2.3 would have detected such a change if used over the final two years of the study period. The data for weeks 1–209 only (essentially the first four years, 1986–1989) were used to derive a model with constant probabilities p_1 and p_2 for the weekly numbers of firearm homicides as a proportion of all homicides. For each r from 210 to 313 the conditional distribution (under this model) of X_r given the full history $\mathbf{X}^{(r-1)}$ was then computed, and the extremeness or otherwise of the observation x_r was assessed with the help of a plot of forecast pseudo-residuals, which is shown as Figure 15.5.

The result is clear. Not one of the 78 weeks 210–287 produces a pseudo-residual segment lying outside the bands at 0.5% and 99.5%. Weeks 288–313 are very different, however. Weeks 297, 298, 299 and 305 all produce segments lying entirely above the 99.5% level, and within weeks 288–313 there are none even close to the 0.5% level. We therefore conclude that, although the data for 1990 and the first half of 1991 are reasonably consistent with a model based on the four years 1986–1989, the data for the second half of 1991 are not; after June 1991, firearm homicides are at a higher level, relative to all homicides, than is consistent with the model.

15.5 Proportion in each of the five categories

As a final illustration of the application to these data of HMMs, we describe here two multinomial–HMMs for the weekly totals in each of the five categories of death. These models are of the kind introduced in Section 8.4.1. Each has two states. One model has constant 'success probabilities' and the other allows for a change in these probabilities at time 287. The model without change-point has ten parameters: two to determine the Markov chain, and four independently determined probabilities for each of the two states. The model with change-point has 18 parameters, since there are eight independent success probabilities relevant to the period before the change-point, and eight after. For the model without change-point, $-l$ is 1810.059, and for the model with change-point it is 1775.610. The corresponding AIC and BIC values are 3640.1 and 3677.6 (without change-point), and 3587.2 and 3654.7 (with). The model with the change-point at time 287 is therefore preferred, and we give it in full here. The underlying Markov chain has transition probability matrix

$$\begin{pmatrix} 0.541 & 0.459 \\ 0.097 & 0.903 \end{pmatrix}$$

and stationary distribution (0.174, 0.826).

Table 15.7 displays, for the period up to the change-point and the period thereafter, the probability of each category of death in state 1 and in state 2, and the corresponding unconditional probabilities. The most noticeable difference between the period before the change-point and the period thereafter is the sharp increase in the unconditional probability of category 1 (firearm homicide), with corresponding decreases in all the other categories.

Clearly the above discussion does not attempt to pursue all the questions of interest arising from these data that may be answered by the fitting of HM (or other) time series models. It is felt, however, that the models described, and the conclusions that may be drawn, are sufficiently illustrative of the technique to make clear its utility in such an application.

Table 15.7 *Multinomial–HMM with change-point at time 287. Probabilities associated with each category of death, before and after the change-point.*

Weeks 1–287

	category 1	2	3	4	5
in state 1	0.124	0.665	0.053	0.098	0.059
in state 2	0.081	0.805	0.024	0.074	0.016
unconditional	0.089	0.780	0.029	0.079	0.023

Weeks 288–313

	category 1	2	3	4	5
in state 1	0.352	0.528	0.010	0.075	0.036
in state 2	0.186	0.733	0.019	0.054	0.008
unconditional	0.215	0.697	0.018	0.058	0.013

Categories: 1   firearm homicide
2   nonfirearm homicide
3   firearm suicide
4   nonfirearm suicide
5   legal intervention homicide.

CHAPTER 16

A model for animal behaviour which incorporates feedback

16.1 Introduction

Animal behaviourists are interested in the causal factors that determine behavioural sequences — i.e. when animals perform particular activities, and under which circumstances they switch to alternative activities. It is accepted that observed behaviour results from the nervous system integrating information regarding the physiological state of the animal, e.g. the levels of nutrients in the blood, with sensory inputs, e.g. concerning the levels of nutrients in a food (Barton Browne, 1993). The combined physiological and perceptual state of the animal is termed the 'motivational state' (McFarland, 1999). MacDonald and Raubenheimer (1995) modelled behaviour sequences using an HMM whose unobserved underlying states were interpreted as motivational states. Their model captures an important aspect of the causal structure of behaviour, since an animal in a given motivational state (e.g. hungry) might perform not only the most likely behaviour for that state (feed) but also other behaviours (groom, drink, walk, etc.). There is not a one-to-one correspondence between motivational state and behaviour. And it is the runlength distributions of the motivational states that are of interest, rather than those of the observed behaviours. HMMs do not, however, take into account the fact that, in many cases, behaviour also influences motivational state; feeding, for example, leads to satiation.

We describe here a model, proposed by Zucchini, Raubenheimer and MacDonald (2008), that incorporates such feedbacks, and we apply it in order to model observed feeding patterns of caterpillars. We define a 'nutrient level', which is determined by the animal's recent feeding behaviour and which, in turn, influences the probability of transition to a different motivational state. Latent-state models, including HMMs, provide a means of grouping two or more behaviours, such as feeding and grooming, into 'activities'. Here the activities of interest are meal-taking, an activity characterized by feeding interspersed by brief pauses, and inter-meal intervals, in which the animal mainly rests but might also feed for brief periods.

The proposed model is not an HMM because the states do not form

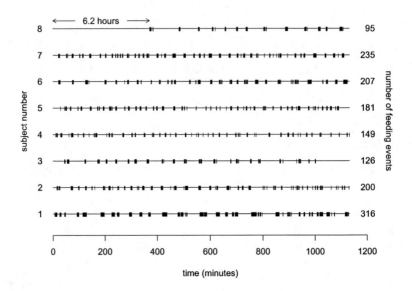

Figure 16.1 *Feeding behaviour of eight* Helicoverpa armigera *caterpillars observed at 1-minute intervals.*

a Markov chain. It can be regarded as a special-purpose extension of HMMs, and shares many of the features of HMMs. We therefore discuss the theoretical aspects of the model before moving on to an application.

In Section 16.9 we demonstrate the application of the model to data collected in an experiment in which eight caterpillars were observed at one-minute intervals for almost 19 hours, and classified as feeding or not feeding. The data are displayed in Figure 16.1.

16.2 The model

Suppose an animal is observed at integer times $t = 1, 2, \ldots, T$, and classified as feeding at time t ($X_t = 1$) or not ($X_t = 0$). We propose the following model.

There are two possible (unobserved) motivational states, provisionally labelled 'hungry' (state 1) and 'sated' (state 2). The state process $\{C_t\}$ is a process in which the transition probabilities are driven by a process $\{N_t\}$ termed the 'nutrient level', which takes values in [0,1]. We assume that N_t is some function of N_{t-1} and X_t. The development that follows is applicable more generally, but we shall restrict our attention to the exponential filter:

$$N_t = \lambda X_t + (1 - \lambda)N_{t-1} \qquad (t = 1, 2, \ldots, T). \qquad (16.1)$$

In biological terms, N_t would correspond approximately with the levels of nutrients in the blood (Simpson and Raubenheimer, 1993). We take the transition probabilities

$$\gamma_{ij}(n_t) = \Pr(C_{t+1} = j \mid C_t = i, N_t = n_t), \quad (t = 1, 2, \ldots, T - 1)$$

to be determined as follows by the nutrient level:

$$\text{logit } \gamma_{11}(n_t) = \alpha_0 + \alpha_1 n_t, \quad \text{logit } \gamma_{22}(n_t) = \beta_0 + \beta_1 n_t; \qquad (16.2)$$

and by the row-sum constraints:

$$\gamma_{11}(n_t) + \gamma_{12}(n_t) = 1, \quad \gamma_{21}(n_t) + \gamma_{22}(n_t) = 1.$$

In Equation (16.2) the logit could be replaced by some other monotonic function $g : (0, 1) \to \mathbb{R}$.

In state 1, the probability of feeding is always π_1, regardless of earlier motivational state, behaviour or nutrient level; similarly π_2 in state 2. The behaviour X_t influences the nutrient level N_t, which in turn determines the transition probabilities of the state process and so influences the state occupied at the next time point, $t + 1$.

Our two fundamental assumptions are as follows, for $t = 2, 3, \ldots, T$ and $t = 1, 2, \ldots, T$ respectively:

$$\Pr(C_t \mid \mathbf{C}^{(t-1)}, N_0, \mathbf{X}^{(t-1)}) = \Pr(C_t \mid C_{t-1}, N_{t-1}); \qquad (16.3)$$

and

$$\begin{aligned} \Pr(X_t = 1 \mid \mathbf{C}^{(t)}, N_0, \mathbf{X}^{(t-1)}) &= \Pr(X_t = 1 \mid C_t) \\ &= \begin{cases} \pi_1 & \text{if } C_t = 1 \\ \pi_2 & \text{if } C_t = 2. \end{cases} \qquad (16.4) \end{aligned}$$

We use the convention that $\mathbf{X}^{(0)}$ is an empty set of random variables, and similarly for $\mathbf{N}^{(0)}$.

Since — given the parameter λ — $\mathbf{N}^{(t-1)}$ is completely determined by N_0 and $\mathbf{X}^{(t-1)}$, the above two assumptions can equivalently be written as

$$\Pr(C_t \mid \mathbf{C}^{(t-1)}, N_0, \mathbf{N}^{(t-1)}, \mathbf{X}^{(t-1)}) = \Pr(C_t \mid C_{t-1}, N_{t-1});$$

and

$$\Pr(X_t = 1 \mid \mathbf{C}^{(t)}, N_0, \mathbf{N}^{(t-1)}, \mathbf{X}^{(t-1)}) = \begin{cases} \pi_1 & \text{if } C_t = 1 \\ \pi_2 & \text{if } C_t = 2. \end{cases}$$

We expect π_1 to be close to 1 and π_2 to be close to 0. If we define

$$p_i(x) = \Pr(X_t = x \mid C_t = i) = \pi_i^x (1 - \pi_i)^{1-x} \quad (\text{for } x = 0, 1; \ i = 1, 2),$$

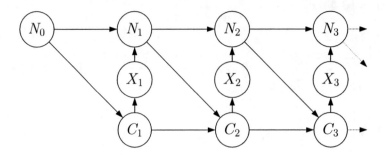

Figure 16.2 *Directed graph representing animal behaviour model.*

assumption (16.4) becomes

$$\Pr(X_t = x \mid \mathbf{C}^{(t)}, N_0, \mathbf{X}^{(t-1)}) = p_{c_t}(x).$$

The model is conveniently represented by the directed graph in Figure 16.2. Notice that there is a path from C_1 to C_3 that does not pass through C_2; the state process is therefore not in general a Markov process.

The following are treated as parameters of the model: α_0, α_1, β_0, β_1, π_1, π_2, and λ. In addition, we take $n_0 \in [0, 1]$ to be a parameter, albeit not one of any intrinsic interest.

One can also treat as a parameter the distribution of C_1 given N_0 (which we denote by the vector $\boldsymbol{\delta}$), or more precisely, the probability $\delta_1 = \Pr(C_1 = 1 \mid N_0)$. It can, however, be shown that on maximizing the resulting likelihood one will simply have either $\hat{\delta}_1 = 1$ or $\hat{\delta}_1 = 0$. We follow this approach, but in the application described in Section 16.9 it turns out in any case that it is biologically reasonable to assume that $\delta_1 = 0$, i.e. that the subject always starts in state 2.

16.3 Likelihood evaluation

Given a sequence of observations $\{\mathbf{x}^{(T)}\}$ assumed to arise from such a model, and given values of the parameters listed above, we need to be able to evaluate the likelihood, both to estimate parameters and to find the marginal and conditional distributions we shall use in this analysis. First we write the likelihood as a T-fold multiple sum, then we show that this sum can be efficiently computed via a recursion.

16.3.1 The likelihood as a multiple sum

We therefore seek

$$L_T = \Pr\left(\mathbf{X}^{(T)} \mid N_0\right) = \sum_{c_1,\ldots,c_T} \Pr\left(\mathbf{C}^{(T)}, \mathbf{X}^{(T)} \mid N_0\right).$$

The summand $\Pr\left(\mathbf{C}^{(T)}, \mathbf{X}^{(T)} \mid N_0\right)$ may be decomposed as follows:

$$\Pr\left(\mathbf{C}^{(T)}, \mathbf{X}^{(T)} \mid N_0\right)$$
$$= \Pr(C_1 \mid N_0) \Pr(X_1 \mid C_1, N_0) \times$$
$$\prod_{t=2}^{T} \left(\Pr(C_t \mid \mathbf{C}^{(t-1)}, N_0, \mathbf{X}^{(t-1)}) \Pr(X_t \mid \mathbf{C}^{(t)}, N_0, \mathbf{X}^{(t-1)}) \right)$$
$$= \Pr(C_1 \mid N_0) \Pr(X_1 \mid C_1) \prod_{t=2}^{T} \left(\Pr(C_t \mid C_{t-1}, N_{t-1}) \Pr(X_t \mid C_t) \right).$$

The first equality follows by repeated application of the definition of conditional probability, and the second from assumptions (16.3) and (16.4). We therefore conclude that

$$L_T = \sum_{c_1,\ldots,c_T} \left(\delta_{c_1}\, p_{c_1}(x_1) \prod_{t=2}^{T} \left(\gamma_{c_{t-1},c_t}(n_{t-1})\, p_{c_t}(x_t) \right) \right).$$

16.3.2 Recursive evaluation

The likelihood is therefore a sum of the form

$$L = \sum_{c_1=1}^{2} \sum_{c_2=1}^{2} \cdots \sum_{c_T=1}^{2} \left(\alpha_1(c_1) \prod_{t=2}^{T} f_t(c_{t-1}, c_t) \right); \qquad (16.5)$$

that is, if here we define

$$\alpha_1(j) = \delta_j\, p_j(x_1)$$

and, for $t = 2, 3, \ldots, T$,

$$f_t(i, j) = \gamma_{ij}(n_{t-1})\, p_j(x_t).$$

Multiple sums of the form (16.5) can in general be evaluated recursively; see Exercise 7 of Chapter 2. Indeed, this is the key property of hidden Markov likelihoods that makes them computationally feasible, and here makes it unnecessary to sum explicitly over the 2^T terms. (See Lange (2002, p. 120) for a very useful discussion of such recursive summation as applied in the computation of likelihoods of pedigrees, a more complex problem than the one we need to consider.)

From Exercise 7 of Chapter 2 (part (b)) we conclude that

$$L = \boldsymbol{\alpha}_1 \mathbf{F}_2 \mathbf{F}_3 \cdots \mathbf{F}_T \mathbf{1}',$$

where the 2×2 matrix $\mathbf{F}_t$ has (i, j) element equal to $f_t(i, j)$, and $\boldsymbol{\alpha}_1$ is the vector with jth element $\alpha_1(j)$. In the present context, $\boldsymbol{\alpha}_1 = \boldsymbol{\delta}\mathbf{P}(x_1)$, $\mathbf{F}_2 = \boldsymbol{\Gamma}(n_1)\mathbf{P}(x_2)$, and similarly $\mathbf{F}_3$, etc. Hence the likelihood of the model under discussion can be written as the matrix product

$$L_T = \boldsymbol{\delta}P(x_1)\boldsymbol{\Gamma}(n_1)\mathbf{P}(x_2)\boldsymbol{\Gamma}(n_2) \cdots \boldsymbol{\Gamma}(n_{T-1})\mathbf{P}(x_T)\mathbf{1}', \qquad (16.6)$$

where the vector $\boldsymbol{\delta}$ is the distribution of C_1 given N_0, $\mathbf{P}(x_t)$ is the diagonal matrix with ith diagonal element $p_i(x_t)$, and $\boldsymbol{\Gamma}(n_t)$ is the matrix with (i, j) element $\gamma_{ij}(n_t)$.

As usual, precautions have to be taken against numerical underflow, but otherwise the matrix product (16.6) can be used as it stands to evaluate the likelihood. The computational effort is linear in T, the length of the series, in spite of the fact that there are 2^T terms in the sum.

16.4 Parameter estimation by maximum likelihood

Estimation may be carried out by direct numerical maximization of the log-likelihood. Since the parameters π_1, π_2, λ, n_0 and δ_1 are constrained to lie between 0 and 1, it is convenient to reparametrize the model in order to avoid these constraints. A variety of methods were used in this work to carry out the optimization and checking: the Nelder–Mead simplex algorithm, simulated annealing and methods of Newton type, as implemented by the $\mathbf{R}$ functions `optim` and `nlm`.

An alternative to direct numerical maximization would be to use the EM algorithm. In this particular model, however, there are no closed-form expressions for the parameter estimates given the complete data, i.e. given the observations plus the states occupied at all times. The M step of the EM algorithm would therefore involve numerical optimization, and it seems circuitous to apply an algorithm which requires numerical optimization in each iteration, instead of only once. Whatever method is used, however, one has to bear in mind that there may well be multiple local optima in the likelihood.

16.5 Model checking

When we have fitted a model to the observed behaviour of an animal, we need to examine the model to assess its suitability. One way of doing so is as follows.

It is a routine calculation to find the forecast distributions under the fitted model, i.e. the distribution of each X_t given the history $\mathbf{X}^{(t-1)}$.

The probabilities are the ratios of two likelihood values:

$$\Pr(X_t \mid \mathbf{X}^{(t-1)}) = \Pr(\mathbf{X}^{(t)})/\Pr(\mathbf{X}^{(t-1)}) = L_t/L_{t-1}.$$

We denote by $\hat{p}_t$ the probability $\Pr(X_t = 1 \mid \mathbf{X}^{(t-1)})$ computed thus under the model. Since the joint probability function of $\mathbf{X}^{(T)}$ factorizes as follows:

$$\Pr(\mathbf{X}^{(T)}) = \Pr(X_1)\Pr(X_2 \mid X_1)\Pr(X_3 \mid \mathbf{X}^{(2)}) \cdots \Pr(X_T \mid \mathbf{X}^{(T-1)}),$$

we have a problem of the following form. There are T binary observations x_t, assumed to be drawn independently, with $\mathrm{E}(x_t) = \hat{p}_t$. We wish to test the null hypothesis $\mathrm{E}(x_t) = \hat{p}_t$ (for all t), or equivalently

$$H_0 : g(\mathrm{E}(x_t)) = g(\hat{p}_t),$$

where g is the logit transform. We consider an alternative hypothesis of the form

$$H_A : g(\mathrm{E}(x_t)) = f(g(\hat{p}_t)),$$

where f is a smoothing spline (see e.g. Hastie and Tibshirani (1990)). Departure of f from the identity function constitutes evidence against the null hypothesis, and a plot of f against the identity function will reveal the nature of the departure.

16.6 Inferring the underlying state

A question that is of interest in many applications of latent-state models is this: what are the states of the latent process (here $\{C_t\}$) that are most likely (under the fitted model) to have given rise to the observation sequence? This is the decoding problem discussed in Section 5.3. More specifically, 'local decoding' of the state at time t refers to the determination of the state i_t which is most likely at that time, i.e.

$$i_t = \underset{i=1,2}{\operatorname{argmax}}\ \Pr(C_t = i \mid \mathbf{X}^{(T)} = \mathbf{x}^{(T)}).$$

In the context of feeding, local decoding might be of interest for determining the specific sets of sensory and metabolic events that distinguish meal-taking from intermeal breaks (Simpson and Raubenheimer, 1993).

In contrast, global decoding refers to the determination of that sequence of states $c_1, c_2, \ldots, c_T$ which maximizes the conditional probability

$$\Pr(\mathbf{C}^{(T)} = \mathbf{c}^{(T)} \mid \mathbf{X}^{(T)} = \mathbf{x}^{(T)});$$

or equivalently the joint probability

$$\Pr(\mathbf{C}^{(T)}, \mathbf{X}^{(T)}) = \delta_{c_1} \prod_{t=2}^{T} \gamma_{c_{t-1},c_t}(n_{t-1}) \prod_{t=1}^{T} p_{c_t}(x_t).$$

Global decoding can be carried out, both here and in other contexts, by means of the Viterbi algorithm: see Section 5.3.2. For the sake of completeness we present the details here, although little is needed beyond that which is in Section 5.3.2. Define

$$\xi_{1i} = \Pr(C_1 = i, X_1 = x_1) = \delta_i\, p_i(x_1),$$

and, for $t = 2, 3, \ldots, T$,

$$\xi_{ti} = \max_{c_1, c_2, \ldots, c_{t-1}} \Pr(\mathbf{C}^{(t-1)} = \mathbf{c}^{(t-1)}, C_t = i, \mathbf{X}^{(T)} = \mathbf{x}^{(T)}).$$

It can then be shown that the probabilities ξ_{tj} satisfy the following recursion, for $t = 2, 3, \ldots, T$:

$$\xi_{tj} = \left(\max_{i=1,2} \left(\xi_{t-1,i}\, \gamma_{ij}(n_{t-1}) \right) \right) p_j(x_t). \tag{16.7}$$

This provides an efficient means of computing the $T \times 2$ matrix of values ξ_{tj}, as the computational effort is linear in T. The required sequence of states $i_1, i_2, \ldots, i_T$ can then be determined recursively from

$$i_T = \operatorname*{argmax}_{i=1,2} \xi_{Ti}$$

and, for $t = T - 1, T - 2, \ldots, 1$, from

$$i_t = \operatorname*{argmax}_{i=1,2} \left(\xi_{ti}\, \gamma_{i,i_{t+1}}(n_t) \right).$$

16.7 Models for a heterogeneous group of subjects

There are several directions in which the model may be extended if that is useful for the application intended. For instance, we may wish to investigate the effects of subject- or time-specific covariates on feeding behaviour, in which case we would need to model the subjects as a group rather than as individuals.

16.7.1 Models assuming some parameters to be constant across subjects

One, fairly extreme, model for a group of I subjects would be to assume that they are independent and that, apart from nuisance parameters, they have the same set of parameters, i.e. the seven parameters α_0, α_1, β_0, β_1, π_1, π_2 and λ are common to the I subjects. The likelihood is in this case just the product of the I individual likelihoods, and is a function of the seven parameters listed, I values of n_0 and (and if they are treated as parameters) I values of δ_1; $7 + 2I$ parameters in all. At the other extreme is the collection of I individual models for the subjects, each of which has its own set of nine parameters and is assumed independent of the other subjects. In this case the comparable number of parameters is $9I$. Intermediate between these two cases are models that assume that some but not all of the seven parameters listed are common to the I subjects. For instance, one might wish to assume that only the probabilities π_1 and π_2 are common to all subjects.

16.7.2 Mixed models

However, in drawing overall conclusions from a group, it may be useful to allow for between-subject variability by some means other than merely permitting parameters to differ between subjects. One way of doing so is to incorporate random effects; another is to use subject-specific covariates.

The incorporation of a single subject-specific random effect into a model for a group of subjects is in principle straightforward; see Altman (2007) for a general discussion of the introduction of random effects into HMMs. For concreteness, suppose that the six parameters α_0, α_1, β_0, β_1, π_1 and π_2 can reasonably be supposed constant across subjects, but not λ. Instead we suppose that λ is generated by some density f with support $[0,1]$.

Conditional on λ, the likelihood of the observations on subject i is

$$L_T(i, \lambda) = \boldsymbol{\delta}\mathbf{P}(x_{i1})\boldsymbol{\Gamma}(n_{i1})\mathbf{P}(x_{i2})\boldsymbol{\Gamma}(n_{i2}) \cdots \boldsymbol{\Gamma}(n_{i,T-1})\mathbf{P}(x_{iT})\mathbf{1}',$$

where $\{x_{it} : t = 1, \ldots, T\}$ is the set of observations on subject i and similarly $\{n_{it}\}$ the set of values of the nutrient level. Unconditionally this likelihood is $\int_0^1 L_T(i, \lambda) f(\lambda) \, d\lambda$, and the likelihood of all I subjects is the product

$$L_T = \prod_{i=1}^{I} \int_0^1 L_T(i, \lambda) f(\lambda) \, d\lambda. \tag{16.8}$$

Each evaluation of L_T therefore requires I numerical integrations, which can be performed in **R** by means of the function `integrate`, but slow the computation down considerably.

Incorporation of more than one random effect could proceed similarly, but would require the specification of the joint distribution of these effects, and the replacement of each of the one-dimensional integrals appearing in Equation (16.8) by a multiple integral. The evaluation of such multiple integrals would of course make the computations even more time-consuming.

16.7.3 Inclusion of covariates

In some applications — although not the application which motivated this study — there may be covariate information available that could help to explain observed behaviour, e.g. dietary differences, or whether subjects are male or female. The important question then to be answered is whether such covariate information can efficiently be incorporated into the likelihood computation. The building-blocks of the likelihood are the transition probabilities and initial distribution of the latent process, and the probabilities π_1 and π_2 of the behaviour of interest in the two states. Any of these probabilities can be allowed to depend on covariates without greatly complicating the likelihood computation.

If we wished to introduce a (possibly time-dependent) covariate y_t into the probabilities π_i, here denoted $\pi_i(y_t)$, we could take logit $\pi_1(y_t)$ to be $a_1 + a_2 y_t$, and similarly logit $\pi_2(y_t)$. The likelihood evaluation would then present no new challenges, although the extra parameters would of course tend to slow down the optimization. If the covariate y_t were instead thought to affect the transition probabilities, we could define logit $\gamma_{11}(n_t, y_t)$ to be $\alpha_0 + \alpha_1 n_t + \alpha_2 y_t$, and similarly logit $\gamma_{22}(n_t, y_t)$.

16.8 Other modifications or extensions

Other potentially useful extensions are to increase the number of latent states above two, and to change the nature of the state-dependent distribution, e.g. to allow for more than two behaviour categories or for a continuous behaviour variable.

16.8.1 Increasing the number of states

If more than two latent states are required in the model, this can be accommodated, e.g. by using at Equation (16.2) a higher-dimensional analogue of the logit transform. Any such increase potentially brings with it a large increase in the number of parameters, however; if there are m states, there are $m^2 - m$ transition probabilities to be specified, and it might be necessary to impose some structure on the transition probabilities in order to reduce the number of parameters.

16.8.2 Changing the nature of the state-dependent distribution

In this work the observations are binary, and therefore so are the state-dependent distributions, i.e. the conditional distributions of an observation given the underlying state. But there might well be more than two behaviour categories in some series of observations, or the observations might be continuous. That would require the use, in the likelihood computation, of a different kind of distribution from the binary distribution used here. That is a simple matter, and indeed that flexibility at observation level is one of the advantages of HM or similar models; almost any kind of data can be accommodated at observation level without greatly complicating the likelihood computation. But since here the observations feed back into the nutrient level N_t, the feedback mechanism would need to be correspondingly modified.

16.9 Application to caterpillar feeding behaviour

16.9.1 Data description and preliminary analysis

The model was applied to sequences of observations of eight final-instar *Helicoverpa armigera* caterpillars, collected in an experiment designed to quantify developmental changes in the pattern of feeding (Raubenheimer and Barton Browne, 2000). The caterpillars were observed continuously for 1132 minutes, during which time they were scanned at one-minute intervals and scored as either feeding or not feeding. In order to isolate developmental changes from environmental effects, individually-housed caterpillars were fed semi-synthetic foods of homogeneous and constant nutrient composition, and the recordings were made under conditions of constant temperature and lighting. The caterpillars were derived from a laboratory culture, and so had similar ancestry and developmental histories. Figure 16.1 displayed the data.

Some remarks can immediately be made. Firstly, despite the uniform conditions of the experiment, there is considerable between-subject variation, both as regards the density of feeding events and the apparent pattern thereof. The runlengths of the feeding events differ between subjects; see e.g. those of subjects 5 and 6. A striking feature is that subject 8 began feeding only 6.2 hours after the start of the experiment. Closer examination of the original recordings made on this subject revealed that it was initially eating its exuvium (moulted skin), a behaviour which has been demonstrated to be of nutritional significance (Mira, 2000). However, since the nutrient composition of the exuvium is very different from that of the synthetic foods, in what follows only subjects 1–7 were included in the analysis. Also noticeable is the fact that subject 3 stopped feeding more than two hours before the end of the experiment. This anomaly became apparent in the model-checking exercise described below.

16.9.2 Parameter estimates and model checking

The first step in the model-fitting process was to fit a model separately to each of the seven subjects, i.e. to estimate for each one the nine parameters α_0, α_1, β_0, β_1, π_1, π_2, λ, n_0 and δ_1. In all cases δ_1 was estimated as zero, i.e. the subject started in state 2. A convincing explanation for this is that, until the post-moult skin has hardened, insects cannot use

Table 16.1 *Parameter estimates and log-likelihood: individual models for subjects 1–7, and mixed model with six common parameters and random effect for λ.*

subj.	$\hat{\alpha}_0$	$\hat{\alpha}_1$	$\hat{\beta}_0$	$\hat{\beta}_1$	$\hat{\pi}_1$	$\hat{\pi}_2$	$\hat{\lambda}$	$\hat{n}_0$	$-\log L$
1	5.807	−11.257	2.283	2.301	0.936	0.000	0.027	0.240	331.991
2	2.231	−5.284	−0.263	21.019	0.913	0.009	0.032	0.150	347.952
3	4.762	−10.124	2.900	15.912	0.794	0.004	0.080	0.740	225.166
4	2.274	−7.779	1.294	16.285	0.900	0.000	0.056	0.018	298.678
5	3.135	−7.271	1.682	10.908	0.911	0.006	0.097	1.000	332.510
6	3.080	−5.231	1.374	13.970	0.880	0.001	0.043	0.246	291.004
7	3.888	−9.057	0.617	13.341	0.976	0.003	0.054	0.375	315.188
									2142.488
mixed model	2.735	−5.328	2.127	7.083	0.919	0.003	$\hat{\mu}=0.055$ $\hat{\sigma}=0.051$		2230.090

their mouthparts and so behave as if they were sated. In what follows we shall take it that all subjects start in state 2, and shall not further treat δ_1 as a parameter requiring estimation.

Table 16.1 displays *inter alia* the parameter estimates for each of subjects 1–7, and the corresponding values of minus the log-likelihood.

We now use subject 1 as illustrative. Figure 16.3 displays the observed feeding behaviour, the underlying motivational state sequence inferred by means of the Viterbi algorithm, and the nutrient level. Figure 16.4 presents an enlarged version of the observed feeding behaviour and inferred motivational state.

A point to note from the figures is the close correspondence between the series of feeding bouts and the inferred states. However, as is demonstrated by Figure 16.4, feeding bouts were interspersed with brief periods of non-feeding which did not break the continuity of the inferred state. The model thus succeeded in the aim of delimiting states according to the probability distributions of behaviours, rather than the occurrence of behaviours *per se*. The nutrient level for subject 1 ranges from about 0.1 to 0.5; for some other subjects the lower bound can reach zero. The parameter λ determines the (exponential) rate at which the nutrient level diminishes in the absence of feeding. The associated half-life is given by $\log(0.5)/\log(1-\lambda)$. Thus the estimated half-life for subject 1 is approximately 25 minutes.

Figure 16.5 displays the transition probabilities for subject 1 as a function of nutrient level. As expected, $\hat{\gamma}_{11}$ decreases with increasing

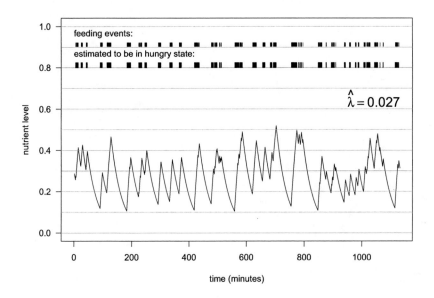

Figure 16.3 *Feeding behaviour, inferred motivational state and nutrient level for subject 1.*

nutrient level, and $\hat{\gamma}_{22}$ increases, and this is true for all seven subjects. Note also that $\hat{\pi}_2$, the estimated probability of feeding when sated, is close to zero. In fact $\hat{\pi}_2$ is less than 0.01 for all seven subjects, and less than 0.001 for two of them. The fact that these estimates are so close to the boundary of the parameter space has implications when one attempts to estimate standard errors. The standard errors of the parameters for subject 1 were estimated by the parametric bootstrap, but are not included here.

Table 16.2 gives runlength statistics, for subjects 1 to 7, of the observed feeding sequences and the estimated sequences of the state 'hungry'. As expected, there are fewer runs for the latter (23% fewer on average) and the mean runlength is larger (45% on average), as is the standard deviation (20% on average).

In applying the model-checking technique of Section 16.5 to subjects 1–7, we were unable to reject H$_0$ in six cases. Using the chi-squared approximation to the distribution of deviance differences, we obtained p-values ranging from 0.30 to 0.82 in these six cases. In the case of subject 3

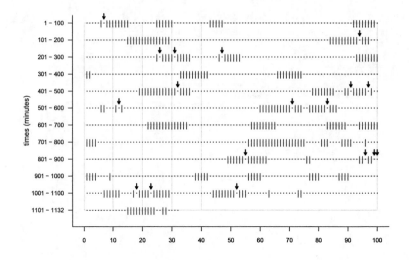

Figure 16.4 *Feeding behaviour of subject 1 (vertical lines), with arrows indicating times at which the subject was not feeding but was inferred to be in state 1, 'hungry'.*

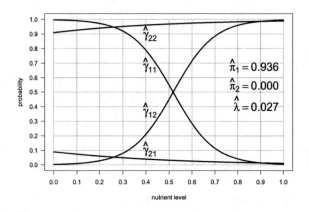

Figure 16.5 *Transition probabilities for subject 1.*

Table 16.2 *Runlength statistics. The seven columns are: subject no.; number of feeding runs, mean length of feeding runs, standard deviation of the length of feeding runs; number of (estimated) hungry runs, mean length of hungry runs, standard deviation of the length of hungry runs.*

subject	feeding runs			estimated hungry runs		
	number	mean	s.d.	number	mean	s.d.
1	58	5.4	4.1	41	8.1	4.9
2	67	3.0	2.3	53	3.9	2.8
3	41	3.1	2.1	22	6.7	2.4
4	57	2.6	1.5	51	3.0	1.7
5	65	2.8	1.6	54	3.5	2.0
6	51	4.1	2.8	35	6.4	3.8
7	57	4.1	2.4	52	4.6	2.7

we concluded that the model fitted is unsatisfactory ($p = 0.051$). Except for subject 3, AIC would also select the hypothesized model.

Examining subject 3 more closely, we note that some of the parameter values are atypical; e.g. the probability of feeding when hungry is atypically low, only about 0.8. The data for subject 3 revealed the unusual feature that there were no feeding events after about time 1000, i.e. no feeding for more than two hours. This is clearly inconsistent with the earlier behaviour of this subject. This conclusion was reinforced by a plot of deviance residuals for subjects 1 and 3, along with spline smooths of these.

16.9.3 Runlength distributions

One of the key questions of biological interest that motivated this work was to assess the extent to which runs of feeding events differ from the runs in motivational state 1 (hungry), and similarly, runs of non-feeding events from runs in state 2 (sated). In an HMM the distributions of the runlengths in the two motivational states are geometric; that cannot be assumed to be the case here. Monte Carlo methods were used here to estimate the runlength distributions under the model.

For each of the seven subjects, a series of length 1 million was generated from the relevant fitted model, and the distribution of each of the four types of run estimated. Figure 16.6 displays plots and summary statistics of the four estimated distributions for subject 1.

The distribution of the feeding runlength clearly differs from that in the hungry state in the expected way. The probability of a runlength being one is almost twice as great for feeding runs as it is for hungry

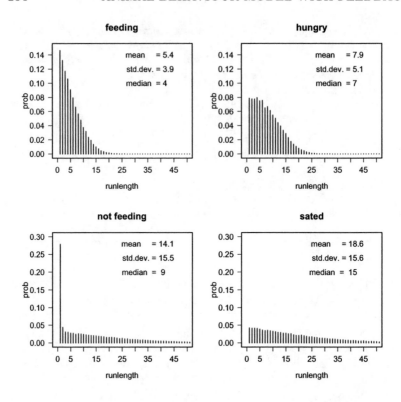

Figure 16.6 *Estimated runlength distributions for subject 1.*

Table 16.3 *Summary of models fitted jointly to the seven subjects. The number of parameters estimated is denoted by k. AIC selects the model with π_1 and π_2 common, and BIC the mixed model, which treats λ as a random effect.*

Model no.	Description	$-\log L$	k	AIC	BIC
1	no parameters common	2142.488	56	4396.976	4787.725
2	π_1 and π_2 common	2153.669	44	**4395.338**	4702.355
3	six parameters common	2219.229	20	4478.458	4618.011
4	seven parameters common	2265.510	14	4559.020	4656.707
5	mixed model (one random effect)	2230.090	15	4490.180	**4594.845**

runs. The mean and median for the latter are greater. The duration of a hungry spell is often longer than one would conclude if one took feeding and hunger as synonymous.

Unlike the distribution of the sated runs, that of the non-feeding runs has a marked peak at 1. This peak is attributable to hungry subjects often interrupting feeding for one time unit, rather than subjects being sated for a single time unit.

A further property worth noting is that the estimated distributions of hungry and sated runlengths do indeed have properties inconsistent with those of geometric distributions (on the positive integers). If μ denotes the mean of such a distribution, the variance is $\mu^2 - \mu$, and the standard deviations corresponding to means of 7.9 and 18.6 would be 7.4 and 18.1; for hungry and sated runs the estimated s.d.s are, however, 5.1 and 15.6 respectively.

16.9.4 Joint models for seven subjects

Five different models were considered for the observations on all seven subjects:

1. a model with no parameters common to the subjects;
2. a model with only π_1 and π_2 common to the subjects;
3. a model with the six parameters α_0, α_1, β_0, β_1, π_1 and π_2 common to the subjects;
4. a model with the seven parameters α_0, α_1, β_0, β_1, π_1, π_2 and λ common to the subjects;
5. a model which incorporated a random effect for the parameter λ and common values for the six parameters α_0, α_1, β_0, β_1, π_1 and π_2. The model used for λ, which is bounded by 0 and 1, was a normal distribution (mean μ and variance σ^2) restricted to the interval [0,1]. This distribution was suggested by a kernel density estimate of λ-estimates from model 3.

In all cases the subjects were taken to be independent of each other and to have started in state 2. As there were only seven subjects, models with more than one random effect were not fitted because one cannot expect to identify a joint distribution of random effects from only seven observations. In the case of model 5, it was necessary to replace a small part of the **R** code by code written in C in order to speed up the computations.

Table 16.3 displays in each case the log-likelihood attained, the number of parameters estimated, AIC and BIC. The model selected by AIC is model 2, which has $\pi_1 = 0.913$ and $\pi_2 = 0.002$ for all subjects. The model selected by BIC is model 5, the mixed model. The parameter estimates for that model appear in Table 16.1, along with those of the seven individual models. An interesting point to note is how much better the mixed model is than model 4. These two models differ only in their treatment of λ; model 4 uses a single fixed value, and the mixed model uses a random effect.

16.10 Discussion

The application of HMMs to animal behaviour (MacDonald and Rauben-heimer, 1995) has hitherto been limited to behaviours whose consequences do not readily alter motivational state.

The model here presented provides an extension which can allow for the important class of behaviours that are feedback-regulated (Toates, 1986). In the present example, the states 'hungry' and 'sated' as used above are of course an artefact of the model, and do not necessarily correspond to the accepted meanings of those terms. A different way of using the model described here would be to define state 1 as the state in which the probability of feeding is (say) 0.9, and similarly state 2 as that with probability (say) 0.1.

Irrespective of how the states are defined, the important point is that their delineation provides an objective means for exploring the physiological and environmental factors that determine the transitions between activities by animals. Such transitions are believed to play an important role in the evolution of behaviour (Sibly and McFarland, 1976), and the understanding of their causal factors is considered a central goal in the study of animal behaviour (Dewsbury, 1992).

The exponential filter — Equation (16.1) — used here for the nutrient level seems plausible. It is, however, by no means the only possibility, and no fundamental difficulty arises if this component of the model is changed. Ideally, the filter should reflect the manner in which feeding affects the motivational process, which might be expected to vary with a range of factors such as the nutrient composition of the foods, the recent feeding history of the animals, their state of health, etc. (Simpson, 1990). In our example, the same filter applied reasonably well across the experimental animals, probably because they were standardized for age and developmental history, and were studied in uniform laboratory conditions. Interestingly, however, a study of foraging by wild grasshoppers revealed that the patterns of feeding were no less regular than those observed in tightly controlled laboratory conditions (Raubenheimer and Bernays, 1993), suggesting that there might be some uniformity of such characteristics as the decay function even in more complex ecological conditions.

The models introduced here, or variants thereof, are potentially of much wider applicability than to feeding behaviour. They may be applied essentially unchanged to any binary behaviour thought to have a feedback effect on some underlying state, and with some modification of the feedback mechanism the restriction to binary observations could be removed.

In overview, by allowing for feedback-regulation we have extended the application of HM or similar models to a wider range of applications in the study of behaviour. We believe that these models hold potential for exploring the relationships among observed behaviours, the activities within which they occur, and the underlying causal factors. Accordingly, they provide a step towards a much-needed objective science of motivation (Kennedy, 1992).

Examples of R code

In this Appendix we give examples of the **R** code used to perform the analyses presented in the book. We do not provide complete coverage of what we have done; only the case of a Poisson–HMM is covered fully, but we also illustrate how that code can be modified to cover two other models. Users are encouraged to experiment with the code and to write their own functions for additional models, starting with simple models, such as the binomial–HMM, and then progressing to more advanced models, e.g. multivariate models or HMMs with covariates. There is, however, other **R** software available which can implement many of our models, e.g. the packages `repeated` (Lindsey, 2008), `msm` (Jackson *et al.*, 2003), and `HiddenMarkov` (Harte, 2008).

We are aware that the code presented in this appendix can be improved. We have sometimes sacrificed efficiency or elegance in the interest of transparency. Our aim has been to give code that is easy to modify in order to deal with alternative models.

The time series is assumed to be stored as a vector x of length n, rather than the T which was used in the body of the text. This is to avoid overwriting the standard **R** abbreviation T for TRUE.

A.1 Fit a stationary Poisson–HMM by direct numerical maximization

Estimation by direct numerical maximization of the likelihood function is illustrated here by code for a stationary Poisson–HMM. The computation uses four functions, which appear in A.1.1– A.1.4:

- `pois.HMM.pn2pw` transforms the natural parameters (the vector of Poisson means λ and the t.p.m. Γ) of an m-state Poisson–HMM into a vector of (unconstrained) working parameters;

- `pois.HMM.pw2pn` performs the inverse transformation;

- `pois.HMM.mllk` computes minus the log-likelihood of a Poisson–HMM for a given set of working parameters; and

- `pois.HMM.mle` estimates the parameters of the model using numerical minimization of minus the log-likelihood.

A.1.1 Transform natural parameters to working

```
1   pois.HMM.pn2pw <- function(m,lambda,gamma)
2   {
3    tlambda  <- log(lambda)
4    tgamma   <- NULL
5    if(m>1)
6      {
7      foo    <- log(gamma/diag(gamma))
8      tgamma<- as.vector(foo[!diag(m)])
9      }
10   parvect <- c(tlambda,tgamma)
11   parvect
12  }
```

The vector **parvect** which is returned contains the working parameters, starting with the entries $\log \lambda_i$ for $i = 1, 2, \ldots, \mathtt{m}$, and followed by τ_{ij} for $i, j = 1, 2, \ldots, \mathtt{m}$ and $i \neq j$. See Section 3.3.1 (pp. 47–49).

A.1.2 Transform working parameters to natural

```
1   pois.HMM.pw2pn <- function(m,parvect)
2   {
3    epar    <- exp(parvect)
4    lambda  <- epar[1:m]
5    gamma   <- diag(m)
6    if(m>1)
7      {
8      gamma[!gamma] <- epar[(m+1):(m*m)]
9      gamma         <- gamma/apply(gamma,1,sum)
10     }
11   delta   <- solve(t(diag(m)-gamma+1),rep(1,m))
12   list(lambda=lambda,gamma=gamma,delta=delta)
13  }
```

The first m entries of **parvect**, the vector of working parameters, are used to compute λ_i, for $i = 1, 2, \ldots, \mathtt{m}$, and the remaining $\mathtt{m}(\mathtt{m} - 1)$ entries to compute $\mathbf{\Gamma}$. The stationary distribution **delta** implied by the matrix **gamma** is also computed and returned; see Exercise 8(a) on p. 26.

A.1.3 Log-likelihood of a stationary Poisson–HMM

```
1   pois.HMM.mllk <- function(parvect,x,m,...)
2   {
3    if(m==1) return(-sum(dpois(x,exp(parvect),log=TRUE)))
4    n         <- length(x)
5    pn        <- pois.HMM.pw2pn(m,parvect)
6    allprobs  <- outer(x,pn$lambda,dpois)
7    allprobs  <- ifelse(!is.na(allprobs),allprobs,1)
8    lscale    <- 0
9    foo       <- pn$delta
```

```
10    for (i in 1:n)
11      {
12      foo      <- foo%*%pn$gamma*allprobs[i,]
13      sumfoo <- sum(foo)
14      lscale <- lscale+log(sumfoo)
15      foo      <- foo/sumfoo
16      }
17    mllk        <- -lscale
18    mllk
19  }
```

This function computes minus the log-likelihood of an m-state model for
a given vector **parvect** of working parameters and a given vector **x** of
observations, some of which may be missing (**NA**). The natural parameters are extracted in line 5. Line 6 computes an n × m array of Poisson
probabilities

$$e^{-\lambda_i}\lambda_i^{x_t}/x_t! \quad (i = 1,\ldots \texttt{m}, \quad t = 1,\ldots,\texttt{n}),$$

line 7 substitutes the value 1 for these probabilities in the case of missing
data, and lines 8–17 implement the algorithm for the log-likelihood given
on p. 47.

A.1.4 ML estimation of a stationary Poisson–HMM

```
1   pois.HMM.mle <- function(x,m,lambda0,gamma0,...)
2   {
3    parvect0  <- pois.HMM.pn2pw(m,lambda0,gamma0)
4    mod       <- nlm(pois.HMM.mllk,parvect0,x=x,m=m)
5    pn        <- pois.HMM.pw2pn(m,mod$estimate)
6    mllk      <- mod$minimum
7    np        <- length(parvect0)
8    AIC       <- 2*(mllk+np)
9    n         <- sum(!is.na(x))
10   BIC       <- 2*mllk+np*log(n)
11   list(lambda=pn$lambda,gamma=pn$gamma,delta=pn$delta,
12             code=pn$code,mllk=mllk,AIC=AIC,BIC=BIC)
13  }
```

This function accepts initial values **lambda0** and **gamma0** for the natural
parameters λ and Γ, converts these to the vector **parvect0** of working parameters, invokes the minimizer **nlm** to minimize –log-likelihood,
and returns parameter estimates and values of selection criteria. The
termination code (**code**), which indicates how **nlm** terminated, is also
returned; see the **R** help for **nlm** for an explanation of the values of
code, and for additional arguments that can be specified to tune the
behaviour of **nlm**.

If a different minimizer, such as **optim**, is used instead of **nlm** it is
necessary to modify the code listed above, to take account of differences
in the calling sequence and the returned values.

A.2 More on Poisson–HMMs, including estimation by EM

A.2.1 Generate a realization of a Poisson–HMM

```
1  pois.HMM.generate_sample <-
2   function(n,m,lambda,gamma,delta=NULL)
3  {
4   if(is.null(delta))delta<-solve(t(diag(m)-gamma+1),rep(1,m))
5   mvect <- 1:m
6   state <- numeric(n)
7   state[1] <- sample(mvect,1,prob=delta)
8   for (i in 2:n)
9     state[i]<-sample(mvect,1,prob=gamma[state[i-1],])
10  x <- rpois(n,lambda=lambda[state])
11  x
12 }
```

This function generates a realization of length n of an m-state HMM with
parameters lambda and gamma. If delta is not supplied, the stationary
distribution is computed (line 4) and used as initial distribution. If delta
is supplied, it is used as initial distribution.

A.2.2 Forward and backward probabilities

```
1  pois.HMM.lalphabeta<-function(x,m,lambda,gamma,delta=NULL)
2  {
3   if(is.null(delta))delta<-solve(t(diag(m)-gamma+1),rep(1,m))
4   n          <- length(x)
5   lalpha     <- lbeta<-matrix(NA,m,n)
6   allprobs   <- outer(x,lambda,dpois)
7   foo        <- delta*allprobs[1,]
8   sumfoo     <- sum(foo)
9   lscale     <- log(sumfoo)
10  foo        <- foo/sumfoo
11  lalpha[,1] <- log(foo)+lscale
12  for (i in 2:n)
13     {
14     foo        <- foo%*%gamma*allprobs[i,]
15     sumfoo     <- sum(foo)
16     lscale     <- lscale+log(sumfoo)
17     foo        <- foo/sumfoo
18     lalpha[,i] <- log(foo)+lscale
19     }
20  lbeta[,n]  <- rep(0,m)
21  foo        <- rep(1/m,m)
22  lscale     <- log(m)
23  for (i in (n-1):1)
24     {
25     foo        <- gamma%*%(allprobs[i+1,]*foo)
26     lbeta[,i]  <- log(foo)+lscale
27     sumfoo     <- sum(foo)
28     foo        <- foo/sumfoo
29     lscale     <- lscale+log(sumfoo)
30     }
31  list(la=lalpha,lb=lbeta)
32 }
```

This function computes the *logarithms* of the forward and backward probabilities as defined by Equations (4.1) and (4.2) on p. 60, in the form of m × n matrices. The initial distribution **delta** is handled as in A.2.1. To reduce the risk of underflow, scaling is applied, in the same way as in A.1.3.

A.2.3 EM estimation of a Poisson–HMM

```
1   pois.HMM.EM <- function(x,m,lambda,gamma,delta,
2                                maxiter=1000,tol=1e-6,...)
3   {
4    lambda.next     <- lambda
5    gamma.next      <- gamma
6    delta.next      <- delta
7    for (iter in 1:maxiter)
8      {
9      lallprobs       <- outer(x,lambda,dpois,log=TRUE)
10     fb   <-   pois.HMM.lalphabeta(x,m,lambda,gamma,delta=delta)
11     la   <-   fb$la
12     lb   <-   fb$lb
13     c    <-   max(la[,n])
14     llk  <- c+log(sum(exp(la[,n]-c)))
15     for (j in 1:m)
16       {
17       for (k in 1:m)
18         {
19         gamma.next[j,k] <- gamma[j,k]*sum(exp(la[j,1:(n-1)]+
20                              lallprobs[2:n,k]+lb[k,2:n]-llk))
21         }
22       lambda.next[j] <- sum(exp(la[j,]+lb[j,]-llk)*x)/
23                            sum(exp(la[j,]+lb[j,]-llk))
24       }
25     gamma.next <- gamma.next/apply(gamma.next,1,sum)
26     delta.next <- exp(la[,1]+lb[,1]-llk)
27     delta.next <- delta.next/sum(delta.next)
28     crit       <- sum(abs(lambda-lambda.next)) +
29                        sum(abs(gamma-gamma.next)) +
30                        sum(abs(delta-delta.next))
31     if(crit<tol)
32       {
33       np       <- m*m+m-1
34       AIC      <- -2*(llk-np)
35       BIC      <- -2*llk+np*log(n)
36       return(list(lambda=lambda,gamma=gamma,delta=delta,
37          mllk=-llk,AIC=AIC,BIC=BIC))
38       }
39     lambda       <- lambda.next
40     gamma        <- gamma.next
41     delta        <- delta.next
42     }
43   print(paste("No convergence after",maxiter,"iterations"))
44   NA
45  }
```

This function implements the EM algorithm as described in Sections 4.2.2 and 4.2.3, with the initial values `lambda`, `gamma` and `delta` for the natural parameters λ, Γ and δ. In each iteration the logs of the forward and backward probabilities are computed (lines 10–12); the forward and backward probabilities themselves would tend to underflow. The log-likelihood l, which is needed in both E and M steps, is computed as follows:

$$l = \log \left(\sum_{i=1}^{m} \alpha_n(i) \right) = c + \log \left(\sum_{i=1}^{m} \exp \left(\log(\alpha_n(i)) - c \right) \right),$$

where c is chosen in such a way as to reduce the chances of underflow in the exponentiation. The convergence criterion `crit` used above is the sum of absolute values of the changes in the parameters in one iteration, and could be replaced by some other criterion chosen by the user.

A.2.4 Viterbi algorithm

```
1   pois.HMM.viterbi <-function(x,m,lambda,gamma,delta=NULL,...)
2   {
3    if(is.null(delta))delta<-solve(t(diag(m)-gamma+1),rep(1,m))
4    n            <- length(x)
5    poisprobs <- outer(x,lambda,dpois)
6    xi           <- matrix(0,n,m)
7    foo          <- delta*poisprobs[1,]
8    xi[1,]     <- foo/sum(foo)
9    for (i in 2:n)
10      {
11      foo      <- apply(xi[i-1,]*gamma,2,max)*poisprobs[i,]
12      xi[i,] <- foo/sum(foo)
13      }
14    iv<-numeric(n)
15    iv[n]       <-which.max(xi[n,])
16    for (i in (n-1):1)
17      iv[i] <- which.max(gamma[,iv[i+1]]*xi[i,])
18    iv
19  }
```

This function computes the most likely sequence of states, given the parameters and the observations, as described on p. 84: see Equations (5.9)–(5.11). The initial distribution `delta` is again handled as in A.2.1.

A.2.5 Conditional state probabilities

```
1   pois.HMM.state_probs <-
2    function(x,m,lambda,gamma,delta=NULL,...)
3   {
4    if(is.null(delta))delta<-solve(t(diag(m)-gamma+1),rep(1,m))
5    n            <- length(x)
```

```
6    fb             <- pois.HMM.lalphabeta(x,m,lambda,gamma,
7                         delta=delta)
8    la             <- fb$la
9    lb             <- fb$lb
10   c              <- max(la[,n])
11   llk            <- c+log(sum(exp(la[,n]-c)))
12   stateprobs <- matrix(NA,ncol=n,nrow=m)
13   for (i in 1:n) stateprobs[,i]<-exp(la[,i]+lb[,i]-llk)
14   stateprobs
15   }
```

This function computes the probability $\Pr(C_t = i \mid \mathbf{X}^{(n)})$ for $t = 1, \ldots n$, $i = 1, \ldots, m$, by means of Equation (5.6) on p. 81. As in A.2.3, the logs of the forward probabilities are shifted before exponentiation by a quantity c chosen to reduce the chances of underflow; see line 11.

A.2.6 Local decoding

```
1    pois.HMM.local_decoding <-
2     function(x,m,lambda,gamma,delta=NULL,...)
3    {
4     stateprobs <-
5         pois.HMM.state_probs(x,m,lambda,gamma,delta=delta)
6     ild <- rep(NA,n)
7     for (i in 1:n) ild[i]<-which.max(stateprobs[,i])
8     ild
9    }
```

This function performs local decoding, i.e. determines for each time t the most likely state, as specified by Equation (5.7) on p. 81.

A.2.7 State prediction

```
1    pois.HMM.state_prediction <-
2     function(x,m,lambda,gamma,delta=NULL,H=1,...)
3    {
4     if(is.null(delta))delta<-solve(t(diag(m)-gamma+1),rep(1,m))
5     n             <- length(x)
6     fb            <- pois.HMM.lalphabeta(x,m,
7                         lambda,gamma,delta=delta)
8     la            <- fb$la
9     c             <- max(la[,n])
10    llk           <- c+log(sum(exp(la[,n]-c)))
11    statepreds <- matrix(NA,ncol=H,nrow=m)
12    foo1          <- exp(la[,n]-llk)
13    foo2          <- diag(m)
14    for (i in 1:H)
15       {
16       foo2           <- foo2%*%gamma
17       statepreds[,i] <- foo1%*%foo2
18       }
19    statepreds
20   }
```

This function computes the probability $\Pr(C_t = i \mid \mathbf{X}^{(n)})$ for $t = n+1, \ldots, n+H$ and $i = 1, \ldots, m$, by means of Equation (5.12) on p. 86. As in A.2.3, the logs of the forward probabilities are shifted before exponentiation by a quantity c chosen to reduce the chances of underflow; see line 10.

A.2.8 Forecast distributions

```
1   pois.HMM.forecast <- function(x,m,lambda,gamma,
2          delta=NULL,xrange=NULL,H=1,...)
3   {
4    if(is.null(delta))
5          delta<-solve(t(diag(m)-gamma+1),rep(1,m))
6    if(is.null(xrange))
7          xrange<-qpois(0.001,min(lambda)):
8                  qpois(0.999,max(lambda))
9    n          <- length(x)
10   allprobs  <- outer(x,lambda,dpois)
11   allprobs  <- ifelse(!is.na(allprobs),allprobs,1)
12   foo        <- delta*allprobs[1,]
13   sumfoo     <- sum(foo)
14   lscale     <- log(sumfoo)
15   foo        <- foo/sumfoo
16   for (i in 2:n)
17     {
18     foo      <- foo%*%gamma*allprobs[i,]
19     sumfoo   <- sum(foo)
20     lscale   <- lscale+log(sumfoo)
21     foo      <- foo/sumfoo
22     }
23   xi         <- matrix(NA,nrow=m,ncol=H)
24   for (i in 1:H)
25     {
26     foo      <- foo%*%gamma
27     xi[,i]   <- foo
28     }
29   allprobs  <- outer(xrange,lambda,dpois)
30   fdists     <- allprobs%*%xi[,1:H]
31   list(xrange=xrange,fdists=fdists)
32   }
```

This function uses Equation (5.4) on p. 77 to compute the forecast distributions $\Pr(X_{n+h} = x \mid \mathbf{X}^{(n)})$ for $h = 1, \ldots, H$ and a range of x values. This range can be specified via **xrange**, but if not, a suitably wide range is used: see lines 7 and 8.

A.2.9 Conditional distribution of one observation given the rest

```
1   pois.HMM.conditionals <-
2     function(x,m,lambda,gamma,delta=NULL,xrange=NULL,...)
3   {
```

```
 4   if(is.null(delta))
 5     delta    <- solve(t(diag(m)-gamma+1),rep(1,m))
 6   if(is.null(xrange))
 7     xrange <-qpois(0.001,min(lambda)):
 8               qpois(0.999,max(lambda))
 9   n        <- length(x)
10   fb       <- pois.HMM.lalphabeta(x,m,lambda,gamma,delta=delta)
11   la       <- fb$la
12   lb       <- fb$lb
13   la       <- cbind(log(delta),la)
14   lafact <- apply(la,2,max)
15   lbfact <- apply(lb,2,max)
16   w        <- matrix(NA,ncol=n,nrow=m)
17   for (i in 1:n)
18     {
19     foo    <- (exp(la[,i]-lafact[i])%*%gamma)*
20                 exp(lb[,i]-lbfact[i])
21     w[,i] <- foo/sum(foo)
22     }
23   allprobs <- outer(xrange,lambda,dpois)
24   cdists   <- allprobs%*%w
25   list(xrange=xrange,cdists=cdists)
26 }
```

This function computes $\Pr(X_t = x \mid \mathbf{X}^{(-t)})$ for a range of x values and $t = 1, \ldots, n$ via Equation (5.3) on p. 77. The range can be specified via **xrange**, but if not, a suitably wide range is used. Here also the logs of the forward and the backward probabilities are shifted before exponentiation by a quantity chosen to reduce the chances of underflow; see lines 19 and 20.

A.2.10 Ordinary pseudo-residuals

```
 1   pois.HMM.pseudo_residuals <-
 2    function(x,m,lambda,gamma, delta=NULL,...)
 3   {
 4    if(is.null(delta))delta<-solve(t(diag(m)-gamma+1),rep(1,m))
 5    n         <- length(x)
 6    cdists    <- pois.HMM.conditionals(x,m,lambda, gamma,
 7                      delta=delta,xrange=0:max(x))$cdists
 8    cumdists <- rbind(rep(0,n),apply(cdists,2,cumsum))
 9    ul <- uh <- rep(NA,n)
10    for (i in 1:n)
11      {
12      ul[i]    <- cumdists[x[i]+1,i]
13      uh[i]    <- cumdists[x[i]+2,i]
14      }
15    um        <- 0.5*(ul+uh)
16    npsr      <- qnorm(rbind(ul,um,uh))
17    npsr
18   }
```

This function computes, for each t from 1 to n, the ordinary normal pseudo-residuals as described in Section 6.2.2. These are returned in an n × 3 matrix in which the columns give (in order) the lower, mid- and upper pseudo-residuals.

A.3 HMM with bivariate normal state-dependent distributions

This section presents the code for fitting an m-state HMM with bivariate normal state-dependent distributions, by means of the discrete likelihood. It is assumed that the observations are available as intervals: x1 and x2 are n × 2 matrices of lower and upper bounds for the observations. (The values -inf and inf are permitted. Note that missing values are thereby allowed for.) The natural parameters are the means mu (an m × 2 matrix), the standard deviations sigma (an m × 2 matrix), the vector corr of m correlations, and the t.p.m. gamma.

There are in all $m^2 + 4m$ parameters to be estimated, and each evaluation of the likelihood requires mn bivariate-normal probabilities of a rectangle: see line 19 in A.3.3. This code must therefore be expected to be slow.

An example of the use of this code appears in Section 10.4.

A.3.1 Transform natural parameters to working

```
1   bivnorm.HMM.pn2pw <-
2   function(mu,sigma,corr,gamma,m)
3   {
4    tsigma <- log(sigma)
5    tcorr<-log((1+corr)/(1-corr))
6    tgamma <- NULL
7    if(m>1)
8      {
9      foo     <- log(gamma/diag(gamma))
10     tgamma <- as.vector(foo[!diag(m)])
11     }
12   parvect <- c(as.vector(mu),as.vector(tsigma),
13               tcorr,tgamma)
14   parvect
15  }
```

The working parameters are assembled in the vector **parvect** in the order: mu, sigma, corr, gamma. The means are untransformed, the s.d.s are log-transformed, the correlations are transformed from $(-1, 1)$ to $\mathbb{R}$ by $\rho \mapsto \log((1 + \rho)/(1 - \rho))$, and the t.p.m. transformed in the usual way, i.e. as in Section 3.3.1 (pp. 47–49).

A.3.2 Transform working parameters to natural

```
1   bivnorm.HMM.pw2pn <- function(parvect,m)
2   {
3    mu       <- matrix(parvect[1:(2*m)],m,2)
4    sigma <- matrix(exp(parvect[(2*m+1):(4*m)]),m,2)
5    temp<-exp(parvect[(4*m+1):(5*m)])
6    corr<-(temp-1)/(temp+1)
7    gamma <- diag(m)
8    if(m>1)
9       {
10      gamma[!gamma]<-exp(parvect[(5*m+1):(m*(m+4))])
11      gamma<-gamma/apply(gamma,1,sum)
12      }
13   delta<-solve(t(diag(m)-gamma+1),rep(1,m))
14   list(mu=mu,sigma=sigma,corr=corr,gamma=gamma,
15         delta=delta)
16  }
```

A.3.3 Discrete log-likelihood

```
1   bivnorm.HMM.mllk<-function(parvect,x1,x2,m,...)
2   {
3    n          <- dim(x1)[1]
4    p          <- bivnorm.HMM.pw2pn(parvect,m)
5    foo        <- p$delta
6    covs       <- array(NA,c(2,2,m))
7    for (j in 1:m)
8       {
9       covs[,,j] <- diag(p$sigma[j,])%*%
10                 matrix(c(1, p$corr[j],p$corr[j],1),2,2)%*%
11                 diag(p$sigma[j,])
12      }
13   P          <- rep(NA,m)
14   lscale     <- 0
15   for (i in 1:n)
16      {
17      for (j in 1:m)
18         {
19         P[j] <- pmvnorm(lower=c(x1[i,1],x2[i,1]),
20                    upper=c(x1[i,2],x2[i,2]),
21                    mean=p$mu[j,], sigma=covs[,,j])
22         }
23      foo     <- foo%*%p$gamma*P
24      sumfoo  <- sum(foo)
25      lscale  <- lscale+log(sumfoo)
26      foo     <- foo/sumfoo
27      }
28   mllk       <- -lscale
29   mllk
30  }
```

Lines 7–12 assemble the covariance matrices in the $2 \times 2 \times m$ array covs.
Lines 14–28 compute minus the log-likelihood. The package mvtnorm is
needed by this function, for the function pmvnorm.

A.3.4 MLEs of the parameters

```
1   bivnorm.HMM.mle<-
2     function(x1,x2,m,mu0,sigma0,corr0,gamma0,...)
3   {
4     n        <- dim(x1)[1]
5     start <- bivnorm.HMM.pn2pw(mu0,sigma0,corr0,gamma0,m)
6     mod      <- nlm(bivnorm.HMM.mllk,p=start,x1=x1,x2=x2,m=m,
7                     steptol = 1e-4,iterlim = 10000)
8     mllk  <- mod$minimum
9     code  <- mod$code
10    p        <- bivnorm.HMM.pw2pn(mod$estimate,m)
11    np       <- m*(m+4)
12    AIC   <- 2*(mllk+np)
13    BIC   <- 2*mllk+np*log(n)
14    list(mu=p$mu,sigma=p$sigma,corr=p$corr,
15              gamma=p$gamma,delta=p$delta,code=code,
16              mllk=mllk,AIC=AIC,BIC=BIC)
17  }
```

A.4 Fitting a categorical HMM by constrained optimization

As a final illustration of the variations of models and estimation techniques that are possible, we describe here how the constrained optimizer constrOptim can be used to fit a categorical HMM, i.e. a model of the kind that is discussed in Section 8.4.2. We assume that the underlying stationary Markov chain has m states, and that associated with each state i there are q probabilities adding to 1: $\sum_{j=1}^{q} \pi_{ij} = 1$. There are m^2 transition probabilities to be estimated, and mq state-dependent probabilities π_{ij}, a total of $m(m+q)$ parameters. The constraints (apart from nonnegativity of all the parameters) are in two groups:

$$\sum_{j=1}^{m} \gamma_{ij} = 1 \quad \text{and} \quad \sum_{j=1}^{q} \pi_{ij} = 1, \quad \text{for } i = 1, 2, \ldots m.$$

However, it is necessary to restructure these $2m$ constraints slightly in order to use constrOptim. We rewrite them as

$$\sum_{j=1}^{m-1} (-\gamma_{ij}) \geq -1 \quad \text{and} \quad \sum_{j=1}^{q-1} (-\pi_{ij}) \geq -1, \quad \text{for } i = 1, 2, \ldots m.$$

In this formulation there are $(m^2 - m) + (mq - m)$ parameters, all subject to nonnegativity constraints, and subject to a further $2m$ constraints; in all, $m(m+q)$ constraints must be supplied to constrOptim.

A model which we have checked by this means is the two-state model for categorized wind direction, described in Section 12.2.1. Our experience in this and other applications leads us to conclude provisionally that `constrOptim` can be very slow compared to transformation and un-constrained maximization by `nlm`. The **R** help for `constrOptim` states that it is likely to be slow if the optimum is on a boundary, and indeed the optimum is on a boundary in the wind-direction application and many of our other models. The resulting parameter estimates differ very little from those supplied by `nlm`.

Note that, if $q = 2$, the code presented here can be used to fit models to binary time series, although more efficient functions can be written for that specific case.

A.4.1 Log-likelihood

```
1   cat.HMM.nllk  <- function(parvect,x,m,q,...)
2   {
3     n        <- length(x)
4     gamma  < -matrix(0,m,m)
5     pr       <- matrix(0,m,q)
6     for (i in 1:m)
7       {
8       gamma[i,1:(m-1)] <-
9         parvect[((i-1)*(m-1)+1):(i*(m-1))]
10      gamma[i,m]<-1-sum(gamma[i,1:(m-1)])
11      pr[i,1:(q-1)] <-
12        parvect[((i-1)*(q-1)+m*(m-1)+1):
13                       (i*(q-1)+m*(m-1))]
14      pr[i,q] <- 1-sum(pr[i,1:(q-1)])
15      }
16    delta    <- solve(t(diag(m)-gamma+1),rep(1,m))
17    lscale <- 0
18    foo      <- delta
19    for (i in 1:n)
20      {
21      foo      <- foo%*%gamma*pr[,x[i]]
22      sumfoo <- sum(foo)
23      lscale <- lscale+log(sumfoo)
24      foo      <- foo/sumfoo
25      }
26    nllk <- -lscale
27    nllk
28    }
```

In lines 4–15 the vector of working parameters `parvect` is unpacked into two matrices of natural parameters, `gamma` and `pr`. The stationary distribution `delta` is computed in line 16, and then minus log-likelihood `nllk` is computed in lines 17–26. The usual scaling method is applied in order to reduce the risk of numerical underflow.

A.4.2 MLEs of the parameters

```
1   cat.HMM.mle <- function(x,m,gamma0,pr0,...)
2   {
3    q          <- ncol(pr0)
4    parvect0 <- c(as.vector(t(gamma0[,-m])),
5                  as.vector(t(pr0[,-q])))
6    np         <- m*(m+q-2)
7    u1         <- diag(np)
8    u2 <- u3 <- matrix(0,m,np)
9    for (i in 1:m)
10      {
11      u2[i,((i-1)*(m-1)+1):(i*(m-1))] <- rep(-1,m-1)
12      u3[i,(m*(m-1)+(i-1)*(q-1)+1):(m*(m-1)+i*(q-1))]
13            <- rep(-1,q-1)
14      }
15    ui   <- rbind(u1,u2,u3)
16    ci   <- c(rep(0,np),rep(-1,2*m))
17    mod <- constrOptim(parvect0,cat.HMM.nllk,grad=NULL,ui=ui,
18            ci=ci,mu=1e-07,method="Nelder-Mead",x=x,m=m,q=q)
19    mod
20  }
```

In lines 6–16 we set up the constraints needed as input to `constrOptim`, which then returns details of the fitted model. The object `mod` contains (among other things) the vector of working parameters `mod$par`: see the **R** help for `constrOptim`.

Some proofs

In this appendix we present proofs of five results, shown here in boxes, that are used principally in Section 4.1. None of these results is surprising given the structure of an HMM. Indeed a more intuitive, and less laborious, way to establish such properties is to invoke the separation properties of the directed graph of the model. More precisely, if one can establish that the sets of random variables $\mathbf{A}$ and $\mathbf{B}$ in a directed graphical model are 'd-separated' by the set $\mathbf{C}$, it will then follow that $\mathbf{A}$ and $\mathbf{B}$ are conditionally independent given $\mathbf{C}$; see Pearl (2000, pp. 16–18) or Bishop (2006, pp. 378 and 619). An account of the properties of HMMs that is similar to the approach we follow here is provided by Koski (2001, Chapter 13). We present the results for the case in which the random variables X_t are discrete. Analogous results hold in the continuous case.

B.1 A factorization needed for the forward probabilities

The first purpose of the appendix is to establish the following result, which we use in Section 4.1.1 in order to interpret $\alpha_t(i)$ as the forward probability $\Pr(\mathbf{X}^{(t)}, C_t = i)$. (Recall that $\alpha_t(i)$ was defined as the ith element of $\boldsymbol{\alpha}_t = \boldsymbol{\delta}\mathbf{P}(x_1)\prod_{s=2}^{t}\boldsymbol{\Gamma}\mathbf{P}(x_s)$: see Equation (2.15).)

For positive integers t:

$$\Pr(\mathbf{X}^{(t+1)}, C_t, C_{t+1}) = \Pr(\mathbf{X}^{(t)}, C_t)\,\Pr(C_{t+1}\mid C_t)\,\Pr(X_{t+1}\mid C_{t+1}).$$
$$\text{(B.1)}$$

Throughout this appendix we assume that

$$\Pr(C_t\mid \mathbf{C}^{(t-1)}) = \Pr(C_t\mid C_{t-1})$$

and

$$\Pr(X_t\mid \mathbf{X}^{(t-1)}, \mathbf{C}^{(t)}) = \Pr(X_t\mid C_t).$$

In addition, we assume that these (and other) conditional probabilities are defined, in which case the probabilities that appear as denominators in what follows are strictly positive. The model may be represented, as usual, by the directed graph in Figure B.1.

The tool we use throughout this appendix is the following factorization for the joint distribution of the set of random variables V_i in a directed

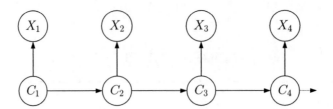

Figure B.1 *Directed graph of basic HMM.*

graphical model, which appeared earlier as Equation (2.5):

$$\Pr(V_1, V_2, \ldots, V_n) = \prod_{i=1}^{n} \Pr(V_i \mid \mathrm{pa}(V_i)), \tag{B.2}$$

where $\mathrm{pa}(V_i)$ denotes all the parents of V_i in the set V_1, V_2, ..., V_n. In our model, the only parent of X_k is C_k, and (for $k = 2, 3, \ldots$) the only parent of C_k is C_{k-1}; C_1 has no parent. The joint distribution of $\mathbf{X}^{(t)}$ and $\mathbf{C}^{(t)}$, for instance, is therefore given by

$$\Pr(\mathbf{X}^{(t)}, \mathbf{C}^{(t)}) = \Pr(C_1) \prod_{k=2}^{t} \Pr(C_k \mid C_{k-1}) \prod_{k=1}^{t} \Pr(X_k \mid C_k). \tag{B.3}$$

In order to prove Equation (B.1), note that Equation (B.3) and the analogous expression for $\Pr(\mathbf{X}^{(t+1)}, \mathbf{C}^{(t+1)})$ imply that

$$\Pr(\mathbf{X}^{(t+1)}, \mathbf{C}^{(t+1)}) = \Pr(C_{t+1} \mid C_t) \, \Pr(X_{t+1} \mid C_{t+1}) \, \Pr(\mathbf{X}^{(t)}, \mathbf{C}^{(t)}).$$

Now sum over $\mathbf{C}^{(t-1)}$; the result is Equation (B.1).   □

Furthermore, (B.1) can be generalized as follows.

For any (integer) $T \geq t + 1$:
$$\Pr(\mathbf{X}_1^T, C_t, C_{t+1}) = \Pr(\mathbf{X}_1^t, C_t) \, \Pr(C_{t+1} \mid C_t) \, \Pr(\mathbf{X}_{t+1}^T \mid C_{t+1}). \tag{B.4}$$

(Recall the notation $\mathbf{X}_a^b = (X_a, X_{a+1}, \ldots, X_b)$.) Briefly, the proof of (B.4) proceeds as follows. First write $\Pr(\mathbf{X}_1^T, \mathbf{C}_1^T)$ as

$$\Pr(C_1) \prod_{k=2}^{T} \Pr(C_k \mid C_{k-1}) \prod_{k=1}^{T} \Pr(X_k \mid C_k),$$

then split each of the two products into $k \leq t$ and $k \geq t+1$. Use the fact that

$$\Pr(\mathbf{X}_{t+1}^T, \mathbf{C}_{t+1}^T) = \Pr(C_{t+1}) \prod_{k=t+2}^T \Pr(C_k \mid C_{k-1}) \prod_{k=t+1}^T \Pr(X_k \mid C_k),$$

and sum $\Pr(\mathbf{X}_1^T, \mathbf{C}_1^T)$ over $\mathbf{C}_{t+2}^T$ and $\mathbf{C}_1^{t-1}$. $\square$

B.2 Two results needed for the backward probabilities

In this section we establish the two results used in Section 4.1.2 in order to interpret $\beta_t(i)$ as the backward probability $\Pr(\mathbf{X}_{t+1}^T \mid C_t = i)$.

The first of these is that,

for $t = 0, 1, \ldots, T-1$,
$$\Pr(\mathbf{X}_{t+1}^T \mid C_{t+1}) = \Pr(X_{t+1} \mid C_{t+1}) \Pr(\mathbf{X}_{t+2}^T \mid C_{t+1}). \qquad \text{(B.5)}$$

This is established by noting that

$$\Pr(\mathbf{X}_{t+1}^T, \mathbf{C}_{t+1}^T)$$

$$= \Pr(X_{t+1} \mid C_{t+1}) \left(\Pr(C_{t+1}) \prod_{k=t+2}^T \Pr(C_k \mid C_{k-1}) \prod_{k=t+2}^T \Pr(X_k \mid C_k) \right)$$

$$= \Pr(X_{t+1} \mid C_{t+1}) \Pr(\mathbf{X}_{t+2}^T, \mathbf{C}_{t+1}^T),$$

and then summing over $\mathbf{C}_{t+2}^T$ and dividing by $\Pr(C_{t+1})$. $\square$

The second result is that,

for $t = 1, 2, \ldots, T-1$,
$$\Pr(\mathbf{X}_{t+1}^T \mid C_{t+1}) = \Pr(\mathbf{X}_{t+1}^T \mid C_t, C_{t+1}). \qquad \text{(B.6)}$$

This we prove as follows. The right-hand side of Equation (B.6) is

$$\frac{1}{\Pr(C_t, C_{t+1})} \sum_{\mathbf{C}_{t+2}^T} \Pr(\mathbf{X}_{t+1}^T, \mathbf{C}_t^T),$$

which by (B.2) reduces to

$$\sum_{\mathbf{C}_{t+2}^T} \prod_{k=t+2}^T \Pr(C_k \mid C_{k-1}) \prod_{k=t+1}^T \Pr(X_k \mid C_k).$$

The left-hand side is

$$\frac{1}{\Pr(C_{t+1})} \sum_{\mathbf{C}_{t+2}^T} \Pr(\mathbf{X}_{t+1}^T, \mathbf{C}_{t+1}^T),$$

which reduces to the same expression. $\square$

B.3 Conditional independence of $\mathbf{X}_1^t$ and $\mathbf{X}_{t+1}^T$

Here we establish the conditional independence of $\mathbf{X}_1^t$ and $\mathbf{X}_{t+1}^T$ given C_t, used in Section 4.1.3 to link the forward and backward probabilities to the probabilities $\Pr(\mathbf{X}^{(T)} = \mathbf{x}^{(T)}, C_t = i)$. That is, we show that,

for $t = 1, 2, \ldots, T - 1$,

$$\Pr(\mathbf{X}_1^T \mid C_t) = \Pr(\mathbf{X}_1^t \mid C_t) \Pr(\mathbf{X}_{t+1}^T \mid C_t). \tag{B.7}$$

To prove this, first note that

$$\Pr(\mathbf{X}_1^T, \mathbf{C}_1^T) = \Pr(\mathbf{X}_1^t, \mathbf{C}_1^t) \frac{1}{\Pr(C_t)} \Pr(\mathbf{X}_{t+1}^T, \mathbf{C}_t^T),$$

which follows by repeated application of Equation (B.2). Then sum over $\mathbf{C}_1^{t-1}$ and $\mathbf{C}_{t+1}^T$. This yields

$$\Pr(\mathbf{X}_1^T, C_t) = \Pr(\mathbf{X}_1^t, C_t) \frac{1}{\Pr(C_t)} \Pr(\mathbf{X}_{t+1}^T, C_t),$$

from which the result is immediate. $\square$

References

Abramowitz, M., Stegun, I.A., Danos, M. and Rafelski, J. (eds.) (1984). *Pocketbook of Mathematical Functions. Abridged Edition of Handbook of Mathematical Functions.* Verlag Harri Deutsch, Thun and Frankfurt am Main.

Aitchison, J. (1982). The statistical analysis of compositional data. *J. Roy. Statist. Soc.* B **44**, 139–177.

Albert, P.S. (1991). A two-state Markov mixture model for a time series of epileptic seizure counts. *Biometrics* **47**, 1371–1381.

Altman, R. MacKay (2007). Mixed hidden Markov models: an extension of the hidden Markov model to the longitudinal data setting. *J. Amer. Statist. Assoc.* **102**, 201–210.

Altman, R. MacKay and Petkau, J.A. (2005). Application of hidden Markov models to multiple sclerosis lesion count data. *Statist. Med.* **24**, 2335–2344.

Aston, J.A.D. and Martin, D.E.K. (2007). Distributions associated with general runs and patterns in hidden Markov models. *Ann. Appl. Statist.* **1**, 585–611.

Azzalini, A. and Bowman, A.W. (1990). A look at some data on the Old Faithful geyser. *Appl. Statist.* **39**, 357–365.

Barton Browne, L. (1993). Physiologically induced changes in resource-oriented behaviour. *Ann. Rev. Entomology* **38**, 1–25.

Baum, L.E. (1972). An inequality and associated maximization technique in statistical estimation for probabilistic functions of Markov processes. In *Proc. Third Symposium on Inequalities*, O. Shisha (ed.), 1–8. Academic Press, New York.

Baum, L.E., Petrie, T., Soules, G. and Weiss, N. (1970). A maximization technique occurring in the statistical analysis of probabilistic functions of Markov chains. *Ann. Math. Statist.* **41**, 164–171.

Bellman, R. (1960). *Introduction to Matrix Analysis.* McGraw-Hill, New York.

Berchtold, A. (1999). The double chain Markov model. *Commun. Stat. Theory Meth.* **28**, 2569–2589.

Berchtold, A. (2001). Estimation in the mixture transition distribution model. *J. Time Series Anal.* **22**, 379–397.

Berchtold, A. and Raftery, A.E. (2002). The mixture transition distribution model for high-order Markov chains and non-Gaussian time series. *Statist. Sci.* **17**, 328–356.

Bisgaard, S. and Travis, L.E. (1991). Existence and uniqueness of the solution of the likelihood equations for binary Markov chains. *Statist. Prob. Letters* **12**, 29–35.

Bishop, C.M. (2006). *Pattern Recognition and Machine Learning*. Springer, New York.

Box, G.E.P., Jenkins, G.M. and Reinsel, G.C. (1994). *Time Series Analysis, Forecasting and Control*, third edition. Prentice Hall, Englewood Cliffs, NJ.

Boys, R.J. and Henderson, D.A. (2004). A Bayesian approach to DNA sequence segmentation (with discussion). *Biometrics* **60**, 573–588.

Brockwell, A.E. (2007). Universal residuals: A multivariate transformation. *Statist. Prob. Letters* **77**, 1473–1478.

Bulla, J. and Berzel, A. (2008). Computational issues in parameter estimation for stationary hidden Markov models. *Computat. Statist.* **23**, 1–18.

Bulla, J. and Bulla, I. (2007). Stylized facts of financial time series and hidden semi-Markov models. *Computat. Statist. & Data Analysis* **51**, 2192–2209.

Calvet, L. and Fisher, A.J. (2001). Forecasting multifractal volatility. *J. Econometrics* **105**, 27–58.

Calvet L. and Fisher, A.J. (2004). How to forecast long-run volatility: regime switching and the estimation of multifractal processes. *J. Financial Econometrics* **2**, 49–83.

Cappé, O., Moulines, E. and Rydén, T. (2005). *Inference in Hidden Markov Models*. Springer, New York.

Celeux, G., Hurn, M. and Robert, C.P. (2000). Computational and inferential difficulties with mixture posterior distributions. *J. Amer. Statist. Assoc.* **95**, 957–970.

Chib, S. (1996). Calculating posterior distributions and modal estimates in Markov mixture models. *J. Econometrics* **75**, 79–97.

Chopin, N. (2007). Inference and model choice for sequentially ordered hidden Markov models. *J. Roy. Statist. Soc.* B **69**, 269–284.

Congdon, P. (2006). Bayesian model choice based on Monte Carlo estimates of posterior model probabilities. *Computat. Statist. & Data Analysis* **50**, 346–357.

Cook, R.D. and Weisberg, S. (1982). *Residuals and Influence in Regression*. Chapman & Hall, London.

Cosslett, S.R. and Lee, L.-F. (1985). Serial correlation in latent discrete variable models. *J. Econometrics* **27**, 79–97.

Cox, D.R. (1981). Statistical analysis of time series: some recent developments. *Scand. J. Statist.* **8**, 93–115.

Cox, D.R. (1990). Role of models in statistical analysis. *Statist. Sci.* **5**, 169–174.

Cox, D.R. and Snell, E.J. (1968). A general definition of residuals (with discussion). *J. Roy. Statist. Soc.* B **30**, 248–275.

Davison, A.C. (2003). *Statistical Models*. Cambridge University Press, Cambridge.

Dempster, A.P., Laird, N.M. and Rubin, D.B. (1977). Maximum likelihood from incomplete data via the EM algorithm (with discussion). *J. Roy. Statist. Soc.* B **39**, 1–38.

Dewsbury, D.A. (1992). On the problems studied in ethology, comparative psychology, and animal behaviour. *Ethology* **92**, 89–107.

Diggle, P.J. (1993). Contribution to the discussion on the meeting on the Gibbs sampler and other Markov chain Monte Carlo methods. *J. Roy. Statist. Soc.* B **55**, 67–68.

Draper, D. (2007). Contribution to the discussion of Raftery *et al.* (2007) (pp. 36–37).

Dunn, P.K. and Smyth, G.K. (1996). Randomized quantile residuals. *J. Comp. Graphical Statist.* **5**, 236–244.

Durbin, R., Eddy, S.R., Krogh, A. and Mitchison, G. (1998). *Biological Sequence Analysis: Probabilistic Models of Proteins and Nucleic Acids.* Cambridge University Press, Cambridge.

Efron, B. and Tibshirani, R.J. (1993). *An Introduction to the Bootstrap.* Chapman & Hall, New York.

Ephraim, Y. and Merhav, N. (2002). Hidden Markov processes. *IEEE Trans. Inform. Th.* **48**, 1518–1569.

Feller, W. (1968). *An Introduction to Probability Theory and Its Applications, Volume 1,* third edition. Wiley, New York.

Fisher, N.I. (1993). *The Analysis of Circular Data.* Cambridge University Press, Cambridge.

Fisher, N.I. and Lee, A.J. (1983). A correlation coefficient for circular data. *Biometrika* **70**, 327–332.

Fisher, N.I. and Lee, A.J. (1994). Time series analysis of circular data. *J. Roy. Statist. Soc.* B **56**, 327–339.

Forney, G.D. (1973). The Viterbi algorithm. *Proc. IEEE* **61**, 268–278.

Franke, J. and Seligmann, T. (1993). Conditional maximum-likelihood estimates for INAR(1) processes and their application to modelling epileptic seizure counts. In *Developments in Time Series Analysis*, T. Subba Rao (ed.), 310–330. Chapman & Hall, London.

Fredkin, D.R. and Rice, J.A. (1992). Bayesian restoration of single-channel patch clamp recordings. *Biometrics* **48**, 427–448.

Fridman, M. and Harris, L. (1998). A maximum likelihood approach for non-Gaussian stochastic volatility models. *J. Bus. Econ. Statist.* **16**, 284–291.

Frühwirth-Schnatter, S. (2006). *Finite Mixture and Markov Switching Models.* Springer, New York.

Gill, P.E., Murray, W., Saunders, M.A. and Wright, M.H. (1986). User's Guide for NPSOL: a Fortran package for nonlinear programming. Report SOL 86-2, Department of Operations Research, Stanford University.

Gill, P.E., Murray, W. and Wright, M.H. (1981). *Practical Optimization.* Academic Press, London.

Gould, S.J. (1997). *The Mismeasure of Man*, revised and expanded edition. Penguin Books, London.

Granger, C.W.J. (1982). Acronyms in time series analysis (ATSA). *J. Time Series Anal.* **3**, 103–107.

Green, P.J. (1995). Reversible jump Markov chain Monte Carlo computation and Bayesian model determination. *Biometrika* **82**, 711–732.

Grimmett, G.R. and Stirzaker, D.R. (2001). *Probability and Random Processes*, third edition. Oxford University Press, Oxford.

Guttorp, P. (1995). *Stochastic Modeling of Scientific Data.* Chapman & Hall, London.

Haines, L.M., Munoz, W.P. and van Gelderen, C.J. (1989). ARIMA modelling of birth data. *J. Appl. Statist.* **16**, 55–67.

Haney, D.J. (1993). Methods for analyzing discrete-time, finite state Markov chains. Ph.D. dissertation, Department of Statistics, Stanford University.

Harte, D. (2008). **R** package 'HiddenMarkov', version 1.2-5. URL http://www.statsresearch.co.nz, accessed 27 July 2008.

Hasselblad, V. (1969). Estimation of finite mixtures of distributions from the exponential family. *J. Amer. Statist. Assoc.* **64**, 1459–1471.

Hastie, T.J. and Tibshirani, R.J. (1990). *Generalized Additive Models.* Chapman & Hall, London.

Hastie, T., Tibshirani, R.J. and Friedman, J. (2001). *The Elements of Statistical Learning: Data Mining, Inference and Prediction.* Springer, New York.

Holzmann, H., Munk, A., Suster, M.L. and Zucchini, W. (2006). Hidden Markov models for circular and linear-circular time series. *Environ. Ecol. Stat.* **13**, 325–347.

Hopkins, A., Davies, P. and Dobson, C. (1985). Mathematical models of patterns of seizures: their use in the evaluation of drugs. *Arch. Neurol.* **42**, 463–467.

Hughes, J.P. (1993). A class of stochastic models for relating synoptic atmospheric patterns to local hydrologic phenomena. Ph.D. dissertation, University of Washington.

Ihaka, R. and Gentleman, R. (1996). R: a language for data analysis and graphics. *J. Comp. Graphical Statist.* **5**, 299–314.

Jackson, C.H., Sharples, L.D., Thompson, S.G., Duffy, S.W. and Couto, E. (2003). Multistate Markov models for disease progression with classification error. *The Statistician* **52**, 193–209.

Jacquier, E., Polson, N.G. and Rossi, P.E. (2004). Bayesian analysis of stochastic volatility models with fat-tails and correlated errors. *J. Econometrics* **122**, 185–212.

Jammalamadaka, S.R. and Sarma, Y.R. (1988). A correlation coefficient for angular variables. In *Statistical Theory and Data Analysis II*, K. Matusita (ed.), 349–364. North Holland, New York.

Jammalamadaka, S.R. and SenGupta, A. (2001). *Topics in Circular Statistics.* World Scientific, Singapore.

Jordan, M.I. (2004). Graphical models. *Statist. Science* **19**, 140–155.

Juang, B.H. and Rabiner, L.R. (1991). Hidden Markov models for speech recognition. *Technometrics* **33**, 251–272.

Kelly, F.P. (1979). *Reversibility and Stochastic Networks.* Wiley, Chichester.

Kennedy, J.S. (1992). *The New Anthropomorphism.* Cambridge University Press, Cambridge.

Kim, S., Shephard, N. and Chib, S. (1998). Stochastic volatility: likelihood inference and comparison with ARCH models. *Rev. Econ. Studies* **65**, 361–393.

Koski, T. (2001). *Hidden Markov Models for Bioinformatics.* Kluwer Academic Publishers, Dordrecht.

Lange, K. (1995). A quasi-Newton acceleration of the EM algorithm. *Statistica Sinica* **5**, 1–18.

Lange, K. (2002). *Mathematical and Statistical Methods for Genetic Analysis*, second edition. Springer, New York.

Lange, K. (2004). *Optimization*. Springer, New York.

Lange, K. and Boehnke, M. (1983). Extensions to pedigree analysis V. Optimal calculation of Mendelian likelihoods. *Hum. Hered.* **33**, 291–301.

Le, N.D., Leroux, B.G. and Puterman, M.L. (1992). Reader reaction: Exact likelihood evaluation in a Markov mixture model for time series of seizure counts. *Biometrics* **48**, 317–323.

Leisch, F. (2004). FlexMix: A general framework for finite mixture models and latent class regression in R. *J. Statistical Software* **11**. http://www.jstatsoft.org/v11/i08/.

Leroux, B.G. and Puterman, M.L. (1992). Maximum-penalized-likelihood estimation for independent and Markov-dependent mixture models. *Biometrics* **48**, 545–558.

Levinson, S.E., Rabiner, L.R. and Sondhi, M.M. (1983). An introduction to the application of the theory of probabilistic functions of a Markov process to automatic speech recognition. *Bell System Tech. J.* **62**, 1035–1074.

Lindsey, J.K. (2004). *Statistical Analysis of Stochastic Processes in Time*. Cambridge University Press, Cambridge.

Lindsey, J.K. (2008). **R** package 'Repeated'.
URL http://popgen.unimaas.nl/~jlindsey/rcode.html, accessed 27 July 2008.

Linhart, H. and Zucchini, W. (1986). *Model Selection*. Wiley, New York.

Little, R.J.A. and Rubin, D.B. (2002). *Statistical Analysis with Missing Data*, second edition. Wiley, New York.

Lloyd, E.H. (1980). *Handbook of Applicable Mathematics, Vol. 2: Probability*. Wiley, New York.

Lystig, T.C. and Hughes, J.P. (2002). Exact computation of the observed information matrix for hidden Markov models. *J. Comp. Graphical Statist.* **11**, 678–689.

McCullagh, P. and Nelder, J.A. (1989). *Generalized Linear Models*, second edition. Chapman & Hall, London.

MacDonald, I.L. and Raubenheimer, D. (1995). Hidden Markov models and animal behaviour. *Biometrical J.* **37**, 701–712.

MacDonald, I.L. and Zucchini, W. (1997). *Hidden Markov and Other Models for Discrete-valued Time Series*. Chapman & Hall, London.

McFarland, D. (1999). *Animal Behaviour: Psychobiology, Ethology and Evolution*, third edition. Longman Scientific and Technical, Harlow.

McLachlan, G.J. and Krishnan, T. (1997). *The EM Algorithm and Extensions*. Wiley, New York.

McLachlan, G.J. and Peel, D. (2000). *Finite Mixture Models*. Wiley, New York.

Mira, A. (2000). Exuviae eating: a nitrogen meal? *J. Insect Physiol.* **46**, 605–610.

Munoz, W.P., Haines, L.M. and van Gelderen, C.J. (1987). An analysis of the maternity data of Edendale Hospital in Natal for the period 1970–1985. Part 1: Trends and seasonality. Internal report, Edendale Hospital.

Newton, M.A., and Raftery, A.E., (1994). Approximate Bayesian inference with the weighted likelihood bootstrap (with discussion). *J. Roy. Statist. Soc.* B **56**, 3–48.

Nicolas, P., Bize, L., Muri, F., Hoebeke, M., Rodolphe, F., Ehrlich, S.D., Prum, B. and Bessières, P. (2002). Mining *Bacillus subtilis* chromosome heterogeneities using hidden Markov models. *Nucleic Acids Res.* **30**, 1418–1426.

Omori, Y., Chib, S., Shephard, N. and Nakajima, J. (2007). Stochastic volatility with leverage: fast and efficient likelihood inference. *J. Econometrics* **140**, 425–449.

Pearl, J. (2000). *Causality: Models, Reasoning and Inference.* Cambridge University Press, Cambridge.

Pegram, G.G.S. (1980). An autoregressive model for multilag Markov chains. *J. Appl. Prob.* **17**, 350–362.

R Development Core Team (2008). *R: A language and environment for statistical computing.* R Foundation for Statistical Computing, Vienna, Austria. ISBN 3-900051-07-0, URL http://www.R-project.org.

Raftery, A.E. (1985a). A model for high-order Markov chains. *J. Roy. Statist. Soc.* B **47**, 528–539.

Raftery, A.E. (1985b). A new model for discrete-valued time series: autocorrelations and extensions. *Rassegna di Metodi Statistici ed Applicazioni* **3–4**, 149–162.

Raftery, A.E., Newton, M.A., Satagopan, J.M. and Krivitsky, P.N. (2007). Estimating the integrated likelihood via posterior simulation using the harmonic mean identity (with discussion). In *Bayesian Statistics 8*, J.M. Bernardo, M.J. Bayarri, J.O. Berger, A.P. Dawid, D. Heckerman, A.F.M. Smith and M. West (eds.), 1–45. Oxford University Press, Oxford.

Raftery, A.E. and Tavaré, S. (1994). Estimation and modelling repeated patterns in high order Markov chains with the mixture transition distribution model. *Appl. Statist.* **43**, 179–199.

Raubenheimer, D. and Barton Browne, L. (2000). Developmental changes in the patterns of feeding in fourth- and fifth-instar *Helicoverpa armigera* caterpillars. *Physiol. Entomology* **25**, 390–399.

Raubenheimer, D. and Bernays, E.A. (1993). Patterns of feeding in the polyphagous grasshopper *Taeniopoda eques*: a field study. *Anim. Behav.* **45**, 153–167.

Richardson, S. and Green, P.J. (1997). On Bayesian analysis of mixtures with an unknown number of components (with discussion). *J. Roy. Statist. Soc.* B **59**, 731–792.

Robert, C.P. and Casella, G. (1999). *Monte Carlo Statistical Methods.* Springer, New York.

Robert, C.P., Rydén, T. and Titterington, D.M. (2000). Bayesian inference in hidden Markov models through the reversible jump Markov chain Monte Carlo method. *J. Roy. Statist. Soc.* B **62**, 57–75.

Robert, C.P. and Titterington, D.M. (1998). Reparameterization strategies for hidden Markov models and Bayesian approaches to maximum likelihood estimation. *Statist. and Computing* **8**, 145–158.

Rosenblatt, M. (1952). Remarks on a multivariate transformation. *Ann. Math. Statist.* **23**, 470–472.

Rossi, A. and Gallo, G.M. (2006). Volatility estimation via hidden Markov models. *J. Empirical Finance* **13**, 203–230.

Rydén, T., Teräsvirta, T. and Åsbrink, S. (1998). Stylized facts of daily returns series and the hidden Markov model. *J. Appl. Econometr.* **13**, 217–244.

Schilling, W. (1947). A frequency distribution represented as the sum of two Poisson distributions. *J. Amer. Statist. Assoc.* **42**, 407–424.

Schimert, J. (1992). A high order hidden Markov model. Ph.D. dissertation, University of Washington.

Scholz, F.W. (2006). Maximum likelihood estimation. In *Encyclopedia of Statistical Sciences*, second edition, S. Kotz, N. Balakrishnan, C.B. Read, B. Vidakovic and N.L. Johnson (eds.), 4629–4639. Wiley, Hoboken, NJ.

Scott, D.W. (1992). *Multivariate Density Estimation: Theory, Practice and Visualization*. Wiley, New York.

Scott, S.L. (2002). Bayesian methods for hidden Markov models: Recursive computing in the 21st century. *J. Amer. Statist. Assoc.* **97**, 337–351.

Scott, S.L., James, G.M. and Sugar, C.A. (2005). Hidden Markov models for longitudinal comparisons. *J. Amer. Statist. Assoc.* **100**, 359–369.

Shephard, N.G. (1996). Statistical aspects of ARCH and stochastic volatility. In *Time Series Models: In econometrics, finance and other fields*, D.R. Cox, D.V. Hinkley and O.E. Barndorff-Nielsen (eds.), 1–67. Chapman & Hall, London.

Sibly, R.M. and McFarland, D. (1976). On the fitness of behaviour sequences. *American Naturalist* **110**, 601–617.

Silverman, B.W. (1985). Some aspects of the spline smoothing approach to nonparametric regression curve fitting (with Discussion). *J. Roy. Statist. Soc.* B **47**, 1–52.

Silverman, B.W. (1986). *Density Estimation for Statistics and Data Analysis*. Chapman & Hall, London.

Simpson, S.J. (1990). The pattern of feeding. In *A Biology of Grasshoppers*, R.F. Chapman and T. Joern (eds.), 73–103. Wiley, New York.

Simpson, S.J. and Raubenheimer, D. (1993). The central role of the haemolymph in the regulation of nutrient intake in insects. *Physiol. Entomology* **18**, 395–403.

Singh, G.B. (2003). Statistical Modeling of DNA Sequences and Patterns. In *Introduction to Bioinformatics: A Theoretical and Practical Approach*, S.A. Krawetz and D.D. Womble (eds.), 357–373. Humana Press, Totowa, NJ.

Smyth, P., Heckerman, D. and Jordan, M.I. (1997). Probabilistic independence networks for hidden Markov probability models. *Neural Computation* **9**, 227–269.

Speed, T.P. (2008). Terence's stuff: my favourite algorithm. *IMS Bulletin* **37(9)**, 14.

Spreij, P. (2001). On the Markov property of a finite hidden Markov chain. *Statist. Prob. Letters* **52**, 279–288.

Stadie, A. (2002). Überprüfung stochastischer Modelle mit Pseudo–Residuen. Ph.D. dissertation, Universität Göttingen.

Suster, M.L. (2000). Neural control of larval locomotion in *Drosophila melanogaster*. Ph.D. thesis, University of Cambridge.

Suster, M.L., Martin, J.R., Sung, C. and Robinow, S. (2003). Targeted expression of tetanus toxin reveals sets of neurons involved in larval locomotion in *Drosophila*. *J. Neurobiology* **55**, 233–246.

Timmermann, A. (2000). Moments of Markov switching models. *J. Econometrics* **96**, 75–111.

Titterington, D.M., Smith, A.F.M. and Makov, U.E. (1985). *Statistical Analysis of Finite Mixture Distributions*. Wiley, New York.

Toates, F. (1986). *Motivational Systems*. Cambridge University Press, Cambridge.

Turner, R. (2008). Direct maximization of the likelihood of a hidden Markov model. *Computat. Statist. & Data Analysis* **52**, 4147–4160.

van Belle, G. (2002). *Statistical Rules of Thumb*. Wiley, New York.

Visser, I., Raijmakers, M.E.J. and Molenaar, P.C.M. (2002). Fitting hidden Markov models to psychological data. *Scientific Programming* **10**, 185–199.

Viterbi, A.J. (1967). Error bounds for convolutional codes and an asymptotically optimal decoding algorithm. *IEEE Trans. Inform. Th.* **13**, 260–269.

Wasserman, L. (2000). Bayesian model selection and model averaging. *J. Math. Psychology* **44**, 92–107.

Weisberg, S. (1985). *Applied Linear Regression*, second edition. Wiley, New York.

Welch, L.R. (2003). Hidden Markov models and the Baum–Welch algorithm. *IEEE Inform. Soc. Newsl.* **53**, pp. 1, 10–13.

Whitaker, L. (1914). On the Poisson law of small numbers. *Biometrika* **10**, 36–71.

Wittmann, B.K., Rurak, D.W. and Taylor, S. (1984). Real-time ultrasound observation of breathing and body movements in foetal lambs from 55 days gestation to term. Abstract presented at the XI Annual Conference, Society for the Study of Foetal Physiology, Oxford.

Yu, J. (2005). On leverage in a stochastic volatility model. *J. Econometrics* **127**, 165–178.

Zeger, S.L. and Qaqish, B. (1988). Markov regression models for time series: a quasi-likelihood approach. *Biometrics* **44**, 1019–1031.

Zucchini, W. (2000). An introduction to model selection. *J. Math. Psychology* **44**, 41–61.

Zucchini, W. and Guttorp, P. (1991). A hidden Markov model for space-time precipitation. *Water Resour. Res.* **27**, 1917–1923.

Zucchini, W. and MacDonald, I.L. (1998). Hidden Markov time series models: some computational issues. In *Computing Science and Statistics* **30**, S. Weisberg (ed.), 157–163. Interface Foundation of North America, Inc., Fairfax Station, VA.

Zucchini, W. and MacDonald, I.L. (1999). Illustrations of the use of pseudo-residuals in assessing the fit of a model. In *Statistical Modelling. Proceedings of the 14th International Workshop on Statistical Modelling, Graz, July 19–23, 1999*, H. Friedl, A. Berghold, G. Kauermann (eds.), 409–416.

Zucchini, W., Raubenheimer, D. and MacDonald, I.L. (2008). Modeling time series of animal behavior by means of a latent-state model with feedback. *Biometrics* **64**, 807–815.

Author index

Subject index